MOUNTAINS
OF SOUTHERN AFRICA

MOUNTAINS
OF SOUTHERN AFRICA

TEXT BY
DAVID BRISTOW

PHOTOGRAPHS BY
CLIVE WARD

C. Struik Publishers, Cape Town

ACKNOWLEDGEMENTS

Many people gave unselfishly of their time and knowledge to help make this book what it is – we hope that for them it has been at least as rewarding as it has been for us. A number of kind people throughout the country also offered hospitality, warm food, hot baths and soft beds during our many months on the move. They are too numerous to mention here but the spirit of the book is partly a mirror of their kindness. We would especially like to thank:

Rob Brand for his early geological ideas and direction; Boet du Plooy, who shared with us his extensive knowledge of the country around Aliwal North; Professor Arthur Fuller of Cape Town University's Geology Department, for his extensive and constructive criticism of the text's geological perspective; Dr James Kitching of the Palaeontology Department at University of the Witwatersrand who led us into the world of fossils, put us in touch with many notable people in the field, and allowed us to photograph his specimens; Rob Nan of New Bethesda, who housed and fed us over a Christmas when there was no room at the inn; Bob Scholes, who was like a ghost third author – his expertise in ecology, geology and mountaineering, his broad general knowledge and crisp literary style are an essential part of this book; and Boet van Heerden, of Doornberg farm, whose hospitality and generosity saved us when our 'horse' broke down.

At Struik Publishers, Peter Borchert gave us guidance when we could no longer see the hills for words and pictures, Leni Martin meticulously edited the words and kept our wandering minds on course, and Walther Votteler made creative use of the photographs in his design.

Clive would particularly like to thank:
Mike Fagan and Ian Harvey for assistance with aerial photography; Keith James, the Bard, who eased his journeys through the Cape Fold mountains; Tseliso Ramakhula of the Lesotho Tourist Board, who gave time and assistance as guide and interpreter in his country's mountains; and Tony Shuttleworth who, at great inconvenience, provided workshop facilities to help keep the show on the road.

David's special acknowledgements go to:
Tracey Hawthorne, without whose word processing, proofreading and company during many long hours at the keyboard the text would not have been completed by deadline – she even smiled when told the entire manuscript would have to be completely rewritten and extended; Rodney, his resourceful computer, who kept him amused with games of hide and seek through memory circuits; and Charlene Smith, whose critical eye and interest in the country's prehistory checked the first draft – her encouragement over the years helped keep him at his desk when the mountains beckoned.

DAVID BRISTOW, CLIVE WARD
JOHANNESBURG, 1985

C. Struik (Pty) Ltd
Struik House, Oswald Pirow Street, Foreshore, Cape Town 8001

First edition 1985
Second impression 1988

House editor: Leni Martin
Design by Walther Votteler, Cape Town
Map and drawings by Anne Westoby, Cape Town
Typeset by McManus Bros (Pty) Ltd, Cape Town
Lithographic reproduction by Unifoto (Pty) Ltd, Cape Town
Printed and bound by Printpak Books, Cape Town

ISBN 0 86977 227 9

1. (Opposite title page) *Spitzkoppe, in Namibia, glows in the afternoon sun.* **2.** (Overleaf) *First light washes the pinnacles and spires of Mnweni in the northern Drakensberg.*

CONTENTS

PREFACE

Collecting material for this book took Clive and me across the length and breadth of the country many times. One grey September morning found us deep in the Karoo, en route from Johannesburg to Cape Town, near a spot on the map called Dwyka. Although it is marked on the map, we had some trouble finding it.

The Dwyka geological series, the first layer of the widespread Karoo Supergroup, is composed of the sludge left behind when ice-age glaciers retreated, some 250 million years ago. We thought a beer and a chat with the locals in Dwyka's public house might lead us to where we could see striations chiselled into the bedrock as the ice slid slowly across it.

On discovering that the spot on the map was no more than a lonely railway siding, we headed for the only farm in mile upon mile of grey scrub. An elderly woman greeted us on the cool stoep. With suddenly awakened apprehension I began to explain our presence, trying to spread a little dust on my city accent. Did she perhaps know of any place nearby where these markings might be found? No, she replied curiously, she had never heard of them, but she would call her son.

A young farmer in overalls came onto the stoep, and again we explained what we were looking for. After an awkward silence he replied doubtfully: Well, he too had never heard of such a thing, but perhaps the place to look for ice would be down in the riverbed (which at the time was dry). Or up in the distant hills, his mother thought.

With rough fingers, the farmer measured out a few centimetres in the air and asked just how thick was the ice we were looking for – an inch, a few inches? No, I answered, feet, many feet thick – but I was not actually expecting to see the ice, only the marks of where it had been. Again he said, he had never heard of such a thing. How long ago was the ice here, he queried. 'Mill . . .' I began, thought better of it and said: 'Many, many years ago', not quite sure of his perspective on creation.

A few moments of bemused silence followed as we all narrowed our eyes to look across the parched landscape, locked in one of the worst droughts in living memory. Ice, many feet thick, here? Even Clive and I began to doubt the possibility of it. The only way he could think of ice doing what we said, he continued, was if sometime in the past the nearby dam had burst during midwinter and the ice on it had flowed down the riverbed.

He was not sure about ice gouging out the lines we spoke of; if, however, we said ants had done it, that he could believe. Yes, here the ants were so numerous and aggressive they could certainly eat grooves into the rock. We thanked him and left, knowing that he would get as much amusement in retelling the story as we have.

In the following chapters we will take you on a journey through some of the oldest mountains on earth. They are not the Himalayas, not the Andes nor the Alps, for geologically these are all youngsters. Ours is an African adventure among mountains that are wrinkled and wise.

A geographical and geological unit, the areas we explore are connected by their fauna, their flora and their rocks, as well as by a continuity that transcends all these factors: the Transvaal, Natal, Lesotho, the Cape and Namibia are threaded together like beads on a mountain chain. If the Drakensberg are the scales on the dragon's back, then the escarpment that follows our coastline is the skeleton of the creature that lies curled up and sleeping.

Our journey begins in the time-worn hills of the Magaliesberg. From there it unwinds in a clockwise spiral around the Transvaal, down Natal, through the eastern Cape to the pivotal mountains of the western Cape and then up into the timeless deserts of Namibia. Our adventures have borne deep friendships: they have also developed in us respect and a love for the places we describe. The best way to experience the mountains is to climb them: we hope this book leads you there.

3. Sandstone bands in the Swartberg resemble gnarled muscles of the forces which formed them.   4. (Overleaf) Drakensberg basalts form the high ridges of the Witteberg in the north-eastern Cape Province.

INTRODUCTION
FIRE AND WATER

When Africa stood up out of the water
And the sky was a muddy brown
And the web-footed Quathlamba
* mountains*
Had either to swim or drown
A spirit of great undertaking
Arose from out of the deep –
And a greater and gentler spirit
Counselled only to sleep.
And as one of the voices fell silent
The other more triumphant grew
And troubled the fertile moonlight
Cast slantwise on the Karoo.

Herman Charles Bosman
Life and Death

Somewhere on an outer limb of a galaxy
we call the Milky Way turns a small,
yellow star. Around it spin, at last count,
nine satellite planets and their faithful
moons. The third planet from the sun is
an attractive blue and white globe called
the earth. But four and a half billion years
ago it was conceived as a fiery, seething
blob, probably the result of the great
thermonuclear explosion that gave birth
to our solar system. By a million years
later it had cooled sufficiently to allow a
thin crust to form on the surface, while
inside the furnace raged on.

Piecing together a brief and simplified
version of the earth's past is beset with
problems, for geological history is the
accumulation of field observation and
theories which at times may be based on
no more than the scantiest of evidence.
Any one theory of a geological event (such
as whether the Witwatersrand gold fields
are the result of ancient sedimentary
deposits or later ascending liquid
solutions) has both proposers and
opposers, and it is hardly possible to
present a complete geological jigsaw
puzzle that will satisfy all. The puzzle
I put forward follows a theory that is a
fairly complete and graphic version of our
tumultuous past, but I have drawn on

5. *Tracers follow the paths of stars across
the vast expanse of a Karoo sky.* **6.** *The
Transvaal Drakensberg, land of flowing water.*
7. *(Overleaf) The Pontoks lie locked in the
fierce grip of the Namib Desert.*

other sources where it seemed
appropriate. If a controversial theory or
prediction has been encountered, I have
tended to include rather than reject those
which help give detail to the puzzle, piece
by piece, millennia by millennia.

The loftiest mountain on earth reaches a
height of nearly nine kilometres, and the
deepest ocean trench is a little deeper.
The wave patterns of earthquakes indicate
that the earth's continental crust is, on
average, about 50 kilometres thick but,
like an iceberg, only a very small
percentage protrudes above sea level to
form landmasses. Made up of a number of
major and minor plates, of which the

continents are a part, the crust rests on a
semi-molten or plastic bed about 3 000
kilometres thick, called the mantle.

The crust on the mantle is like a slag on
the surface of a furnace, seething with all
the wonderful impurities of life that have
evolved from the earliest algae. Below the
mantle is an unknown mixed zone, and in
the centre of this a core of very dense
nickel and iron.

The shape of landforms on the crust is
constantly changing. Volcanic activity
brings to the surface new material which
is acted upon by water and wind and
gradually eroded away. Erosion in one
place leads to deposition of the same
material in another, resulting in a
continual redistribution of the surface
materials. In southern Africa periods of
sedimentation and erosion are usually
followed by periods of volcanic activity as
pressure on the mantle is exerted and
released, causing cracks in the crust
through which the boiling lava
periodically bursts.

Until fairly recently, in geological
terms, there was no Africa. The original
landmass, known as Pangaea, split into
two, with today's southern continents and
India making up Gondwanaland. This
giant continent was subjected to tectonic
plate movement from early Palaeozoic
times (500 million years ago) as the plates
of the crust shifted slowly on the tides of
the plastic mantle.

In southern Africa one effect of this
plate movement was the formation of the
Cape Folded Mountain Belt that runs
parallel to the south and west coasts, and
today shows evidence of dramatic
shearing and buckling. One geologist has
even suggested that the subcontinent
south of the Limpopo Mobile Belt, one of
the longest-lived belts of crustal
weakness, could break away from Africa
and form a new island.

The first rock strata, known as the
Primitive System of southern Africa, were

mixture of lavas, sediments and ironstones that were crumpled, torn and metamorphosed by the intrusion of granites. After a long period of erosion the Dominion Group was created, mainly through volcanic activity that formed mountains in the Transvaal and Botswana. Following another long period of erosion by water, sedimentary rocks of the Witwatersrand Supergroup were deposited. During the subsequent volcanic period of the Ventersdorp Supergroup, the ridges laid down during Witwatersrand times were faulted and tilted.

The surface of the land we know as southern Africa was slowly peneplained, or levelled, and the entire area submerged under the ocean. During this submergence the sandy beds and dolomitic limestones of the Transvaal were laid down. With re-emergence came a glacial period, which was followed yet again by erosion and volcanic activity. Internal stresses allowed the intrusion of the Great Dyke of Zimbabwe with its mineral-rich magmas, and later those of the Bushveld Igneous Complex. The area of the subcontinent was uplifted, again peneplained and into a series of shallow inland basins were deposited the Loskop, Matsap and Waterberg sandstones, conglomerates and shales.

Those billions of years are known as the Precambrian and early Palaeozoic eras – a period which slowly settled from great instability into an environment stable enough to allow the earliest single-cell organisms to develop into higher life forms. First developed algae and then primitive plants took shape. In the sea, marine invertebrates evolved into fish, the fish into amphibians and these, as they moved onto land, into reptiles and finally birds and mammals.

A long, relatively dry period of erosion and then sedimentary deposition followed. The Cape rocks are believed to have been laid down mainly as marine deposits. Initially the ocean lay far beyond the present line of the continent, but as the waters encroached northward they deposited a spine of Table Mountain Sandstone along what became the southern rim of Africa. On top of the Table Mountain Group were laid sandstones and shales of the Bokkeveld Group, which are rich in shallow marine fossils including trilobites and brachiopods. Towards the end of the deposition cycle these were replaced by plant fossils. Finally the Witteberg Group was also laid down under oceanic conditions and a number of early ray-finned fish fossils are the most significant

finds, although ten genera and 16 species of vascular plant have also been described from the Witteberg rocks.

These three groups of rock dominate the Cape Fold mountains, from Clanwilliam to Port Alfred and again in Natal. The Cape Supergroup is cut by a number of fault lines such as the Worcester and Cango faults. In September 1969 an earthquake along the Worcester Fault, measuring 6,3 on the Richter scale, caused considerable damage in Ceres and Tulbagh.

Next, the Karoo Supergroup was moulded during successive dynasties of ice, sea, lake, swamp, desert and volcanic fire. Nearly two thirds of South Africa's surface is covered by the horizontal sediments of Karoo rock. The Dwyka ice age that followed the Witteberg Group saw an ice cap extending from the equator's present position far into the Southern Ocean. In fact, evidence suggests that during that period the South Pole was situated somewhere in the area covered by southern Africa today.

During the Cape Supergroup period and while the first Karoo beds were being deposited, the area that was to become Africa heaved itself out of the sea, unleashing the forces that buckled and twisted the rocks of the Cape Supergroup.

The Karoo sediments were finally capped by the Stormberg lavas that formed the Lebombo mountains of Zululand, large surface areas of Namibia and, most spectacularly, the Drakensberg, the youngest and therefore highest mountains in southern Africa.

Intense faulting during the subsequent Cretaceous period, about 100 million years ago, further developed the Cape Fold mountains, and continental uplifting continued as Africa slowly emerged as a single entity. One more brief period of volcanic activity produced the diamond-studded pipes and fissures of kimberlite and, with that, the landscape of southern Africa more or less as we see it today, was complete.

In the north-east and north-west most of the soft Karoo rocks have been eroded away, exposing ancient granites in the Barberton Mountainland and the sediments of the Soutpansberg, Magaliesberg and Waterberg. Shallow pockets of various sedimentary layers are still trapped between folds of older bedrock and give us an idea of how the land has changed its shape and texture.

Pushed up in one spot, the earth's surface soon bulges, folds or bursts open at another. Erosion is a constant force that tries to level off all the protruding areas on

9 10

8. *Afternoon cloud billows on the solid Black and Tan Wall in the Drakensberg.*
9. *Born of fire: quartz crystals are created from sand by the furnace temperatures of the earth's core.*   **10.** *The kokerboom, an aloe species, thrives under the unrelenting blaze of the desert sun.*

11. (Previous page) *Fantastically shaped quartzites of Cedarberg Kloof in the north-western Transvaal are reminiscent of the sculptured rocks of the Cape range.* **12.** *The Wolfberg Cracks in the Cedarberg have been carved by the winds and rain that lash the western Cape in winter.*

the crust. Water, in the form of rivers and rain, and sometimes ice and snow, is the primary agent that eats slowly into the mountains, although wind, too, plays a part. The Himalaya range, the highest in the world, is still being formed as the plate we see as India forces its way under the plate on which Asia lies. Mount Everest and its neighbouring giants have yet to reach unknown heights, but millions of years from now they will be pebbles on the beaches of India.

In some areas, however, the process of erosion has been speeded up to a critical degree. Overgrazing by domestic herds, deforestation and the denuding of natural vegetation allows soil to be carried away at a rate far exceeding that at which it is formed. Mountain slopes are especially susceptible to this washing away of the precious topsoil.

In southern Africa the Drakensberg is the most important rain catchment area and even though large sections are under the control of the Department of Forestry and the Natal Parks Board, much of it is threatened by recreational developments and homeland underdevelopment. South Africa's largest export, topsoil is

irreplaceable and closely linked to our water supply. As it disappears forever into the sea, so the ground's ability to retain water diminishes and the land becomes locked in an escalating cycle of erosion, flood, denudation and aridity.

Whatever the climatic and other influences of the surrounding lower areas, the peaks of Africa form an archipelago of temperate vegetation. As the climatic tides ebb and flow, so the high-altitude species advance down and retreat up their mountain shores. Through the ages the Afro-montane representatives of neighbouring peaks have occasionally been linked and have mixed, then been isolated once more. Under the prevailing climatic conditions they cling to their rocky islands like hardy survivors of a shipwreck. From the genetic material they have rescued, they develop the attributes that best equip them to survive in their new environment and so, over time, evolve into new species.

Those species with the capacity for long distance dispersal are carried far and wide by the winds. Very occasionally one of these genetic life-rafts lands on another island, invigorating the plant life there

with a new species, which can then continue to skip up or down the island 'stepping stones'.

For centuries mountains have held a peculiar attraction for the human animal. They seem to have a spirit or energy field around them that draws wanderers and mountaineers with whispered promises of adventure. Much of that adventure takes the form of climbing which, therefore, plays an integral part in this book. Rock climbing is an adventure that relies on technical ability, to some extent on technology, but mainly on sheer nerve.

Climbers talk of experiencing 'exposure', and understanding what they mean is central to understanding their ethos. Exposure is a refined and rarefied form of fear, the courting and overcoming of which is the essence of climbing. The word contains the seed of its hidden meaning – one 'exposes' oneself to risk. The common element in all climbing, be it on Everest or Table Mountain, exposure is the result of a number of factors, such as height above the ground, openness, where there are no sheltering cracks or comfortable corners, and steepness. Featureless faces, sharp arêtes and overhangs are its natural habitat.

Climbs are graded from A to H according to steepness, difficulty and the commitment needed to lead a route. Grading may seem superfluous but is especially necessary in extreme climbing where each movement is a finely choreographed act of precision balancing and manoeuvring. As walking grades, A and B are seldom used; from C onwards the use of all four limbs is necessary; D grade signifies steep and perhaps exposed scrambling where the use of ropes is recommended; from E grade most climbs are near-vertical but the holds, foot placements and protections are obvious, though strenuous. An experienced climber may decide to 'solo' climb (that is, without ropes or other protection) an E or even higher grade route.

From E grade onwards, each grade is subdivided into three levels of difficulty, thus E1, E2 and E3. F grade is vertical and requires a good level of technical skill for the movements are strenuous and the holds unobvious. This grade marks a watershed of ability: while F3 means very exposed and severe climbing, a climber attempting anything above G1 would have to have made a definite mental and physical commitment to his craft.

Thirty years ago only a handful of climbers could pit themselves against an ascent of G standard. By ten years ago modern safety equipment had made rock climbing far less hazardous and allowed

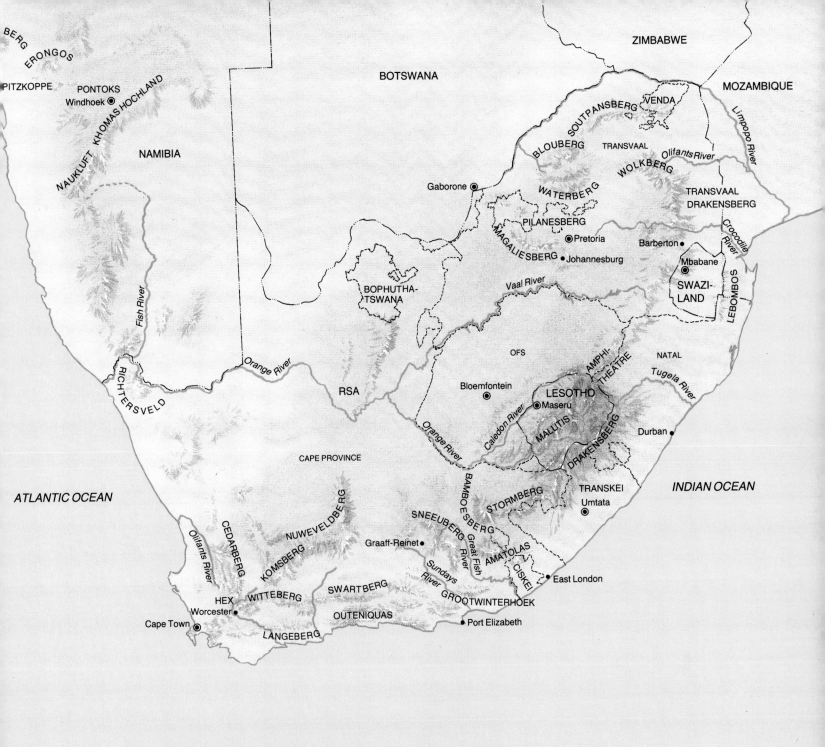

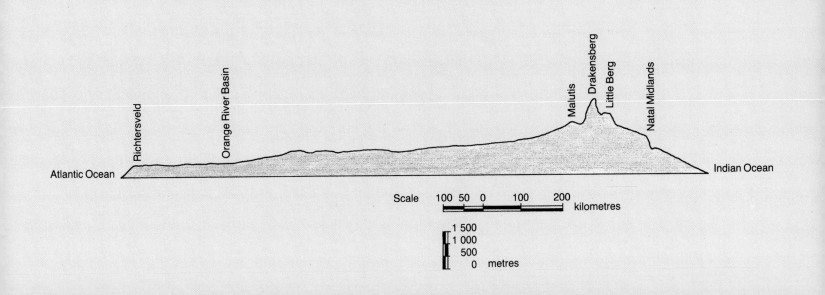

BERG
ERONGOS
PITZKOPPE
PONTOKS
Windhoek

NAUKLUFT KHOMAS HOCHLAND

NAMIBIA

Fish River

RICHTERSVELD

ATLANTIC OCEAN

Orange River

RSA

CAPE PROVINCE

Olifants River

CEDARBERG

NUWEVELDBERG

KOMSBERG

Graaff-Reinet

SNEEUBERG

HEX

WITTEBERG

SWARTBERG

OUTENIQUAS

Worcester

LANGEBERG

Cape Town

Sundays River

GROOTWINTERHOEK

Port Elizabeth

BOTSWANA

ZIMBABWE

MOZAMBIQUE

VENDA

Limpopo River

SOUTPANSBERG

BLOUBERG

TRANSVAAL

WOLKBERG

Olifants River

Gaborone

WATERBERG

TRANSVAAL
DRAKENSBERG

PILANESBERG

MAGALIESBERG

Pretoria

Barberton

Johannesburg

Crocodile River

Mbabane

BOPHUTHA-
TSWANA

Vaal River

SWAZI-
LAND

SOBOMBO

LEBOMBO

OFS

NATAL

AMPHI-
THEATRE

Bloemfontein

Tugela River

LESOTHO

Maseru

Caledon River

MALUTIS

Durban

Orange River

DRAKENSBERG

BAMBOESBERG

STORMBERG

Great Fish River

TRANSKEI

Umtata

INDIAN OCEAN

AMATOLAS

CISKEI

East London

Richtersveld

Atlantic Ocean

Orange River Basin

Malutis

Drakensberg

Little Berg

Natal Midlands

Indian Ocean

Scale   100 50 0   100   200

kilometres

1 500
1 000
500
0   metres

13. *Chalk is used to give climbers' fingers better grip on polished, slippery handholds.*
14. *Years of contact with abrasive rock takes its toll on 'EBs', trusted boots of the rock gymnast. These boots are designed for friction climbing and maintaining an 'edging' grip on rock faces.* **15.** *'Ropes and slings and nuts and things' comprise the basic gear of a modern rock climber. The helmet protects against falling rocks, 'EB' boots and the bag of chalk promote better grip, and the aluminium 'nuts', 'friends' and karrabiner snap clips are placed in cracks where they provide protection for the leader of a climb.* **16.** *Alan Lambert leads 'Migraine' in Tonquani Kloof.* **17.** *Noel Margetts overcomes the exposure of leading a climb, using undercut handholds on a 'thin' face.* **18.** *Alvin Wood places a 'friend' on a difficult G1 route on the Lady's Slipper near Port Elizabeth. The twin cams of the friend are pushed together, placed in a crack and released so that they spring open and jam. A nylon sling and karrabiner snap clip are attached to the friend and the leader passes the rope, tied to the harness around his waist, through the 'karrab'. As he climbs his belayer feeds out the other end of the rope from below, and should the leader slip, the belayer will arrest the fall.*
19. *Horizontal cracks allow Keith James to traverse along a rail on 'Atomic Aloe' in Wolwe River gorge in the Outeniquas.*
20. *(Overleaf) The buckled Attaquas mountains, a folded range bordering the Little Karoo.*

13

14

15

16

the limits to be pushed back with a diminished risk of injury in a fall. Today there are again a handful of climbers capable of H grade climbing and most serious young climbers routinely tackle G routes.

It is the leader of a climbing party who is most exposed and takes most of the risks. In the early days of mountaineering the only equipment used consisted of hobnailed mountaineering boots or rubber-soled sandshoes, and hessian ropes. The leader set off up a face while his partner belayed him by feeding out the rope that was tied to both their waists. As he climbed he passed the rope around or over projecting rocks and if he fell, his 'second' would try to arrest the fall. When he reached a good ledge within his rope's limit he made a secure stance and belayed up the next climber. The second climber was not hauled up but rather the slack rope pulled in to prevent a long plunge in the event of his slipping as he climbed.

Over the past 50 years many sophisticated devices have been invented to make climbing safer. First were iron pitons that were driven into the rock. A short rope sling was tied to the piton and a karrabiner (a metal clip like a large safety pin) at the end of it was clipped around the leader's rope. This is called a running belay, or 'runner'. If the leader fell and was held by his second, he dropped only twice the distance between himself and his last runner.

Today the basic technique remains the same but the use of pitons is generally frowned upon as they have been replaced by aluminium wedges called 'nuts' and

18

19

provide preplaced protection. Nuts and friends are placed by the leader in convenient cracks in the rock and then removed by his second as he moves up to the next belay stance. When opening a route the leader sometimes finds no way of moving further and in such a predicament uses a piton or nut, with a nylon sling to pull himself up on. Referred to as mechanical or aided climbing, this is employed only as a last resort.

The distance between two belay stances, the pitch, is limited only by the length of a modern climbing rope – 45 metres. When a leader is unable to establish a belay by the time his rope runs out and he is unable to retreat, his initiative is tested to the full. Often a pitch is much shorter, combining a series of moves which form a natural and satisfying unit. Abseiling, the method used to descend quickly on a fixed rope, relies on applying friction on the rope, which is attached in such a way that it can be pulled down after the climber has descended safely. Ascending a fixed rope, when climbing the rock is not the objective, is known as prussicking. Named after the German who invented it, this method is an essential part of the 'siege' tactics of expedition climbing. Obviously the leader has to climb the rock or ice first in order to fix the rope.

Even with the latest hardware, rubber climbing boots and nylon ropes there is real danger in climbing, and a climber will do anything in his or her power to prevent a fall. Two friends were recently on a serious climb when the leader, Terry Hoy, shouted down to his belayer, Terry White, that he was about to fall. 'Don't,' shouted back White. 'Your runner's come out.' At the penultimate moment no philosophical debate was needed – a fall would bring death. Hoy managed to stay on the rock and finish the climb.

'Technology in climbing is both a blessing and a curse,' says one of America's top rock climbers, Royal Robbins. 'It expands the limits of the possible but robs us of adventure.' In his book on advanced rock craft he writes: 'A first ascent involves artistic creation in several ways. . . the control, cunning and craft used by the leader in getting up – that is, his artful use of body, of holds, of means of resting and saving his strength and his choice of runners and skill in placing them and making them stay. This area of the leader's self-expression is like a dance which he choreographs as he goes. It doesn't last but he can create another every time he touches rock. And so can all climbers.'

'friends' which do not damage or mark the rock. The evolution of climbing aids is an interesting saga that includes the invention of such esoteric devices as 'chocks', 'stoppers', 'wedgies' and 'panic buttons'. For the specialized climber there are also 'bathooks', 'skyhooks' and 'copperheads'. The story of climbing

evolution can be compared with *Homo habilis* inventing the stone axe; perhaps the next evolutionary step will be *Homo ascendens*.

One of the ethics of climbing has always been purity. Pitons scar the rock and, because they are permanent, not only leave an obvious line to follow, but also

THE
NORTH-WESTERN TRANSVAAL
ANCIENT SCULPTURES

I shall know peace there once again
Where Thebes its mighty ramparts rears
Silent above the desert plain
Wherein the Mara disappears.
Where every morning from the krantz
The eagle spiralling upward flies
Into the glowing vault of heaven
To greet the sun with its shrill cries;
Where sadly, still, hyenas' howl
Wanders through kloof, and echoes loud...

Eugène Marais
Waar Tebes in die Stil Woestyn

The story of the north-western Transvaal mountains is the story of Africa: pastoral herders, raging *impis*, ancient spirits and a wildness, a wideness with which few other places can compare. Regarded by many travellers and trekking parties as a hindrance to their northward and eastward progress, for some these ranges were a place of refuge. They lie in a wide band between the Limpopo and the Olifants rivers, with the most southerly, the Magaliesberg, along the southern limit of the Bushveld Igneous Complex. To the north-west rises the volcanic intrusion of the Pilanesberg, then in a north-eastward sweep looms the Waterberg's bulk, followed at a distance by the Blouberg, an isolated mountain block that forms part of the same geological system as the Soutpansberg, the last of the sequence.

At the close of the eighteenth century, when the south-western Cape was a thriving colonial outpost of Europe, the rest of southern Africa was still considered the dark continent of noble savages and mysteries. Enlightenment took hold only in the 1850s, when David Livingstone ventured on his now-famous expeditions into the interior, and by then some of the subcontinent's most turbulent times were already history.

21. *A lone tree perches above the black cleft of Fernkloof, in the Magaliesberg.* **22.** *Ripple marks on a quartzite block indicate that the Magaliesberg rocks were originally laid down in shallow water as sandstone. Wind and extreme temperatures have carved these characteristic rock sculptures, known as the Twins, near Cedarberg Kloof.*

During the first three decades of the nineteenth century explorers such as Andrew Geddes Bain and Captain William Cornwallis Harris made contact with the many tribes of the hinterland and discovered the vast riches of the land beyond the Vaal River. To them it must have seemed an inexhaustible storehouse of game, minerals, intrigue and adventure. When Cornwallis Harris traversed a range known to the local people as the Kashan

mountains he found herds of game such as had only been dreamed of. Animals in their millions roamed these hot Bushveld plains. He even noted a species not yet recorded – the majestic sable antelope with its gracefully curving scimitar horns and glistening black hide.

The Kashan mountains are now called the Magaliesberg and it is believed they were named after the Tswana chief Magali (Mohale) who ruled a tribe to the north of the mountains. At the time of the white man's early ventures into southern Africa these mountains were the centre of an apocalyptic turmoil which destroyed a way of life that had persisted there for hundreds of years. The relatively peaceful pastoralists who had grown their crops, tended their herds and hunted there since the beginning of their tribal memories, were wiped out and replaced by a bloody tyranny still remembered with awe.

Yet before the arrival of white man, before brown men or even yellow men, indeed before any member of our species walked these hills, they had been a cradle of life. Man's early upright cousin *Australopithecus africanus* lived in the limestone caves of Sterkfontein and Kromdraai. If there is a 'missing link' between man and apes, then *Australopithecus* is as close to it as anthropologists have found, and it was in the dolomite caves near the Magaliesberg that early man developed into a toolmaker, and further. As far back as 250 000 years ago Stone Age man lived in the kloofs and caves here, feeding on the fruits and animals that abounded.

From about 400 AD an Iron Age culture settled in the area, although the Late Stone Age culture, represented by the Bushmen, retained a precarious foothold until relatively recent times. The newcomers from the north, generally

23. *'A world apart' describes the summit of Blouberg, a huge upthrust block of red sandstone that stands isolated on the flat Bushveld of the north-western Transvaal.* **24.** *Rock climbers in Skeperskloof go to great lengths to get onto the highest kloof walls in the province. Crocodile footprints can still be seen on the sandbanks in the lower kloof.* **25.** (Overleaf) *Krantzberg in the Waterberg range, nesting place of vultures and eagles and a favourite haunt of rock climbers, is a series of crags that forms the south-western barrier to the Palala plateau.*

believed to be Bantu-speaking Sotho, prospered as a developing agrarian culture until the last century, when devastating waves of breakaway Zulu *impis* left the area depopulated for the spreading white settlers to annex as their own.

Towards the end of the eighteenth century Dingiswayo, chief of the Mthethwa tribe from the land south of the Black Umfolozi River, journeyed to the Cape Colony. So impressed was he by the military order of the white troops he saw and some other black tribes he met, that he returned to his own country to organize his *impis* into a disciplined and fierce army. He launched brutal attacks on the neighbouring clans and began to establish himself as a tyrant and feared enemy. By 1818 Dingiswayo was recognized as paramount chief of all the clans south of the Black Umfolozi. His bravest and most trusted captain was young Shaka of the Zulu clan, who was soon installed as its chief.

If Dingiswayo's armies were brutal, they must have seemed tame when Shaka unleashed his own armies on the clans not yet under Dingiswayo's influence. In 1822, after Dingiswayo's death, Shaka usurped the crown of the paramount chief and installed himself as king of the Zulu nation. He expanded his dominion south to the Tugela River and sent his own favourite captain, Mzilikazi, to subdue the Sotho tribes far to the north-west.

On returning with his spoils of cattle and slaves, Mzilikazi defied Shaka by keeping the bounty and attempting to set up his own dynasty as chief of the Khumalo, or 'elephant' clan. Fears of an attack from the Zulu, however, drove 'the Bull Elephant' across the Pongola River with his tribe. He crossed the hills near Majuba and first settled in the area where Middelburg now stands.

From here began the path of blood that Mzilikazi's *impis* left behind them as the tribe consolidated into the Matabele nation and moved westwards across the Highveld. With single-minded ruthlessness he crushed all those who stood up to his armies, enslaved the young women and drafted the young men into his forces. He also demanded absolute subservience from even his closest *indunas*. Soon he reigned supreme over an area bounded by the Soutpansberg in the north, the Kalahari in the west, the Vaal River in the south, and the Zulu domain in the east.

The first biting highveld winter led Mzilikazi westward in search of a kinder climate. He built his royal kraal on the banks of the Apies River and then moved it even further west, to the outer edge of the crescent formed where the Magaliesberg range arcs to the north. From here he firmly established himself as a ruler to rival the great Shaka. So fierce were Mzilikazi's raids against the neighbouring tribes that when white explorers from the south ventured into this area, they found the veld littered with skeletons and charred bodies. The remnants of once flourishing tribes were living in trees like terrified primates, harassed by wild beasts and emaciated with hunger.

Robert Moffat, the charismatic missionary who worked among the marauding Griqua and Korana tribes in the northern Cape Province, travelled from his mission station at Kuruman to visit Mzilikazi. Despite being appalled by the horrors of the tyrant's ruthless raids, he was at once impressed by Mzilikazi's youth and friendliness. After Moffat's first reluctant visit to the Magaliesberg a firm friendship was established between himself and the Matabele king: indeed, Mzilikazi considered Moffat's judgement impeccable and considered him an invaluable advisor.

From the Magaliesberg Mzilikazi moved his headquarters to the Marico district. Determined to keep all foreigners out of his territory, he and his marauding *impis* slaughtered a number of advancing Boer trek parties that ventured across the Vaal. This was to be his undoing, for the

oughty pioneers, led by Hendrik otgieter and Gert Maritz, successfully arded off another attack at Vegkop, and a series of battles and skirmishes in the te 1830s and early 1840s routed the tire Matabele nation. His armies in sarray, Mzilikazi fled north of the mpopo River, later re-establishing his ngdom at Bulawayo.

In the mid-nineteenth century, once e dominance of the Matabele in the ransvaal had been broken, the undance of game attracted more and ore hunters, followed by white settlers. did not take long for the vast herds to iminish as wanton killing continued. To oot a giraffe for only one strip of its neck

hide to make a trek whip was commonplace. A Magaliesberg farmer named Retief recalls how as a young man at the turn of this century he helped shoot some of the last surviving big game in the area. Today at his favourite hunting spot, the ford across the Crocodile River in the shadow of the surrounding hills, stands the Pelindaba nuclear research station. The name means 'the meeting is over'.

The Boers who settled in the valleys around the Magaliesberg and Waterberg turned them into one of the most productive farming areas the country had known. Apricots, peaches and citrus fruits from here fetched high prices on the Kimberley diamond fields and, even after

the long wagon trip, a Magaliesberg orange was an expensive treat at the coast. The legendary Oom Schalk Lourens of Herman Charles Bosman's *Marico Tales* savoured Magaliesberg tobacco – reputedly the finest in the land.

The Second Anglo-Boer War brought its own pressures to bear on the area. It was the scene of many campaigns and, during the final stages, a focus of action when the Boers launched guerilla raids to the south from the Bushveld between the Magaliesberg and Waterberg ranges. They used secret pathways to cross the Magaliesberg and at places such as Trident Kloof hikers may still come upon the remains of concealed Boer camps.

Some of the British blockhouses that were built to restrict the movements of the commandos still stand. As part of their plan to counter Boer resistance, these same blockhouses, built from the Magaliesberg's age-old rock, helped enforce the scorched earth policy of the British, when Boer farms were razed to the ground. The blockhouses remain, sterile intrusions of a short-lived dynasty, but once more the area produces peaches, oranges, flowers, cattle and a variety of vegetables to feed the hungry machine of nearby industry and commerce.

Situated between Johannesburg, Pretoria, Krugersdorp and Rustenburg, the Magaliesberg is particularly vulnerable to the demands of a large, industrialized population. Farming, already well established up its slopes, continues to threaten the natural habitat, while stone quarries and silica mines scar the landscape, roads bisect it and numerous towers and masts perch upon it. With the additional attraction of the Hartbeespoort Dam, the range offers one of the few outdoor recreational areas in the southern Transvaal and is thus exposed to even more pressure, particularly from careless picnickers who leave their trail of plastic, beer cans, broken trees and blackened fire places.

The first kloof in the Magaliesberg to be visited by climbers was deep Tonquani, named after Tenquaan, chief of the Bakhatla tribe that took refuge here from Mzilikazi. Harry Barker, a founder member of the Transvaal section of the MCSA (Mountain Club of South Africa) and Dick Barry, whom we shall encounter again in the Drakensberg, opened some of the best climbing routes in Tonquani. Barker, now in his seventies, still climbs where the spirit of 'Tiger' Barry lingers. On the fiftieth anniversary of the opening climb in Tonquani, and accompanied by Harry, I attempted the route 'Feng's Folly' in the miserable rain of early autumn 1984. The well-trodden paths of the popular kloofs must now seem strange to a man who saw leopard, lynx and their cubs melting into the tangled bush before him, and buck scattering at his approach.

Harry Barker was responsible for giving most of the kloofs their modern names, as it was he and his companions who in the 1930s explored the mazes and discovered this climbing mecca. Retief's, Grobelaar's and some others were named after the farmers on whose land they lay. Mhlabatini Kloof means 'the place of white sands', while Trident was the obvious name for a kloof with a triple fork. Aptly enough, it was Barker who in 1968 finally traced the last kloof in the range and named it Hidden Kloof.

Geologically a typical Transvaal range, the Magaliesberg has an African wildness that makes it especially alluring to those, like Barker, who have discovered some of its secrets. These age-old hills remember when the earth was young . . . Walkers in the range may stop to consider a polished rock surface – flat, yet marked with parallel waves that are reminiscent of gently lapping water. The Magaliesberg's quartzites bear the imprint of turbulent geological times, when layer upon layer of sand was deposited on the edges of a large inland waterway. As the Gondwanaland interior dried up to form deserts of

26. 'Hanglip Frontal Super Direct' follows a route straight up the mountain's skyline. It was in the wooded ravines of this area that Eugène Marais carried out his famous observations of baboon behaviour.
27. Brian Gross leads Dave Cheesmond up 'Armageddon' (G1, M2) on Krantzberg. Although this route, opened in the 1960s, remains a classic of its time, a number of G2 and G3 routes have recently been put up the cliffs nearby.

nimaginable magnitude, so sandstone eds were left behind. Millions of years ter the earth's crust split open to emit an oze of molten, mineral-rich volcanic aterial that covered the area north of the andstone rim, tilting the rock upwards long its southern edge. We now call this aucer of volcanic rock the Bushveld gneous Complex – one of the world's ichest mineral deposits.

The tremendous weight of the deep ediments and the heat of the volcanic lows compressed and baked the andstone beds, metamorphosing them into the glazed quartzites we see today. In some areas the temperature and pressure were so great they left a smooth, glazed surface, while the powerful forces which gave the Magaliesberg its characteristic tilt also caused the fissures that guide the orientation of the kloofs.

A natural barrier between the temperate highveld grasslands and the hotter Bushveld, this undulating chain of hills forms a limit to many of the plant and animal species of the two zones, blending them into a unique mountain ecosystem. Marula trees will grow no further south,

the sable will not roam beyond these hills nor the deadly black mamba slither, nor the plum-coloured starling fly. The plant and animal life of the range is thus remarkably diverse for such a confined area: in all groups of organisms excepting fish it harbours more species than the entire British Isles. In one isolated study area 750 plant, 355 vertebrate, including 22 snake, and more than 5 000 insect species have been recorded.

In the dripping moss-green kloofs the Magaliesberg's tranquillity is deepened by the amplified gurgling of streams and falls; the gushing of water quickly absorbs any other noises. Tall trees filter the sunlight and ferns flourish in dark recesses of the dappled ground: roots of the rock fig are strung tautly down dry rock surfaces like harp strings, or bunch along cracks like bulging, muscular limbs. Tonquani, Waterkloof, Fernkloof, Castle Gorge and Grootkloof – the music of these names intimates the range's deceptive beauty.

Soon after swinging northwards to encircle Rustenburg, the Magaliesberg peters out near Boekenhoutkloof, where stands the homestead of Paul Kruger, first president of the Zuid-Afrikaansche Republiek. About 50 kilometres north of this, out of the flat surrounding Bushveld, rises one of the most unusual geological formations in southern Africa. The Pilanesberg forms a circular enclosure, a *laager* of concentric mountain rings that encloses one of the most tranquil and scenic retreats in the area.

During the time of volcanic activity when the Bushveld Igneous formations were laid down, molten lava bubbled up in the area, with a number of small volcanoes forming a cauldron of magma around a fault line. Repeated welling up and subsidence of the lava resulted in a series of ring dykes around the collapsed centre of the cauldron. The original rock cover of these ring dykes has now been almost entirely eroded away and only fragments of the alkaline lavas and coarse breccias remain. However, the Pilanesberg still forms the largest known alkaline complex, rich in countless minerals and a geologist's playground among the volcanic and crystalline formations.

Satellite photographs clearly show the circular cluster of hills, faulted diagonally along the diameter of the circle from south-east to north-west. On the fault line squats Sun City, a large pleasure dome where sun worshippers loll on manicured lawns or in smoky casinos only a hillside away from the rugged African bush.

The republic of Bophuthatswana's

incipal wildlife sanctuary, the
lanesberg National Park, lies next to
n City and is well stocked with game.
ere a policy of selective hunting helps
ntrol game populations and brings in
venue. While hunting in a game reserve
ight appear yet another modern lunacy,
is sound wildlife management in
serves of limited size and few predators,
d helps pay for the importation of other
ecies once indigenous to the area.
From a black eagle's nest on a ledge at
.e reserve's boundary, the view to the
orth-east stretches across the flat
oringbokvlakte, a broad expanse that is
inctuated only by an occasional granite
itcrop before terminating on the far
orizon in a long band of indistinct
imples. These represent the Waterberg,
here the cool mountain streams must
ave seemed like oases to the Boer
ettlers who first crossed the hot plain
get there.

The range tends east-west, with the
ain area lying north of Thabazimbi and
Varmbaths. To the south and east it
onsists of a series of ridges and
scarpments. The central region forms the
alala plateau or Middleveld and to the
xtreme north-east the isolated mountain
our of Blouberg reaches out like a long
m trying to draw the Waterberg towards
ie more distant Soutpansberg.

The rocks of the Waterberg are red,
urple and brown, with iron-rich
ematite as the characteristic pigment.
his sequence of red beds was laid down
etween 1 800 and 1 900 million years
go and is the oldest such sequence
nown. Greenish shales and lighter-
oloured quartz-bearing rocks also occur
n this mass of striking krantzes and
eep valleys.

North-east of Thabazimbi stands the
Krantzberg massif, its steep scree slopes
nd cliffs up to 200 metres high
lominating the landscape. The south-
acing cliffs shelter a large colony of Cape
vultures that for some years has been the
ubject of intense study. Since it has been
orotected the colony has spread along the
edges of Krantzberg and is slowly
annexing many of the fine, high-grade
rock climbing routes along the crag.
Climbers are generally known to be a
conservation-conscious group, but they
also tend to resent being deprived of good
climbing area.

Dick Barry and Harry Barker were the
pioneering cragsmen of the Waterberg too,
with some routes that in their day were
the zenith of climbing achievement. East
of Krantzberg, at Hanglip, they opened the
'Zimbabwe Tower' route in 1937. On the
following day Barry, with Chris Purdham,

28. *Millions of years of weathering has created deep, narrow ravines, giving the summit of Blouberg a maze-like appearance.* **29.** *A Cape vulture chick prepares for its first flight. The vultures' continued existence in the Transvaal mountains is threatened by man's encroachment.* **30.** *A potholed watercourse is one of the natural features that helps to create Blouberg's unique atmosphere.*

opened 'Hanglip Frontal' less than a
month before his last climb. These two
climbs are still known as great all-day
(and frequently all-night!) routes.

From the farmlands below Krantzberg,
steep scree slopes dotted with hardy
highveld proteas rise to the base of the
cliffs, where yellowwood forest spills out
of sheltered gorges. The top of the range is
a rocky terrain of coarse grasses and scrub,
with scattered clusters of silver protea
shrubs and the occasional common
poison-bush.

Among the undulating hills south of
Krantzberg lies a mountain that for
centuries was the source of an Iron Age
culture which dominated the Highveld
until it was torn asunder by Mzilikazi's
conquering *impis*. Metal from
Thabazimbi, the mountain of iron, was
forged by local Sotho tribes, and the
remnants of their foundries and artefacts
may still be seen in many parts of the
Transvaal and Natal. Today Thabazimbi is
still an important iron-mining area,
bringing industry and a railway far into
this otherwise sparsely populated
hinterland.

Soon after the Iron Age communities
were devastated, Voortrekkers began to
penetrate the Waterberg. The first to reach
the central Waterberg was Nicklaas van

Heerden in 1859, and he and his followers
established what is still a conservative
community, with simple farming and
God-fearing ways. Isolation has always
been a feature of these mountains where
rugged landscapes hide lonely
farmsteads; even the motorcar and
telephone have done little to alter the
way of life in the deeper mountain and
plateau areas.

And it was to the Waterberg that one of
the country's most gifted and prodigal
sons fled in search of solitude. Here
Eugène Nielen Marais sought refuge from
his strained social life in Pretoria and
spent his most rewarding years, despite
the continual suffering caused by his
addiction to morphine.

For about nine years he roamed the area
as prospector, hunter, doctor, magistrate,
naturalist and poet. Although he had no
formal medical training he sustained a
life-long interest in medicine and surgery
and was known locally as the miracle
doctor of the Waterberg. Marais was also
renowned for his hypnotic powers over
animal and human alike and this helped
enhance his supernatural reputation.
Much of his time was spent on the farm
Rietfontein in a little house he called the
Ark, encircled by naboom and
boekenhout trees. In these mountains he

31

32

witnessed a nation being reborn after the bitter ravages of war, drought and famine.

During the years of plenty that characterized the first decade of this century, Marais and his friends went on hunting expeditions into the Palala vastness. They came back with trophies of wildebeest, buffalo, kudu, eland, gemsbok and sable, as well as leopard. This was the last great hunting ground of South Africa, remote and wild, a region of plateaux, deep ravines and high krantzes. In the kloofs he encountered troops of baboons which he befriended and observed over many years. These studies led to his well-known books *My Friends the Baboons*

and *The Soul of the Ape*. 'If one loves nature,' he wrote, 'one learns automatically.'

The climatic wheel turned slowly to the dry years of 1911 and 1912 and Marais used this opportunity to study the natural effects of drought. His observations focussed on the complex social organization of the African white ant and in his book *The Soul of the White Ant* he likened their cement-domed colonies to individual living organisms, complete with digestive and circulatory systems and even a primordial intelligence.

Other writings inspired by this land of mystery and imagination were *The Road*

to *Waterberg* and *Dwaalstories*, as well as the poem *Waar Tebes in die Stil Woestyn*, in which Marais describes the flats that stretch between the Waterberg and the Soutpansberg as a 'desert plain'.

Between the two ranges, on this plain, lies Blouberg – the mountain of Maleboch. From the thorn-woodlands of the north-western Transvaal this massif rises in three distinct tiers with great cliffs facing east and north. Part of the sequence of red beds, the mountain is a huge block of land thrust upward between parallel faults. The cliffs are red sandstones and quartzites while the forested or grassy plateaux are composed of laval rubble.

33

Cool mists wash the tall donjons of rock that crowd the sky, waves of green forest lap their bases, and pools on the summit of this mountain island provide welcome sources of water.

Past the sentinel crags stands the highest point of the outcrop, Ga-Monaa-Sena-Moriri – the man with no hair – and from the top of his ever-watchful, patient head one gazes down on the chiefdom of Hananwa. The Hananwa, whose name refers to those who run away (and live to fight another day), are ruled by Chief Maleboch, named after his father and his father's father before him.

From its mountain stronghold the tribe has successfully resisted many regimes and survives as one of the few truly mountain peoples of southern Africa. At sunrise and sunset the Blouberg defies its name, coined from afar in the blue haze, to glow a rosy and then ox-blood red above the green Bushveld. And indeed the mountain has drunk the blood of Hananwa warriors, and womenfolk too.

Chief Maleboch's refusal to recognize the laws and taxes of the Zuid-Afrikaansche Republiek led to the Raadsaal declaring war on his tribe on 12 May 1894. A long siege followed, with the Hananwa braced in a well-stocked cave below Ga-Monaa-Sena-Moriri.

31. *Alvin Wood and Clive Ward wait patiently on a belay ledge while Alan Lambert leads the opening ascent on the crux pitch of 'Gnasher' (G1) in the Magaliesberg.*   **32.** *The north face of Blouberg is the 'big wall' to Transvaal climbers.*   **33.** *'Hourglass' (G1) in lower Tonquani Kloof. These kloof crags were the scene of the first serious rock climbing in the province.*

Dynamite, 200 pounds of cayenne pepper and the Staatsartillerie failed to dislodge the defiant defenders, but eventually thirst drove them out after the women, who sneaked water to the besieged by night, had all been sniped by State marksmen. These defiant defenders turned out to be no more than a handful of starved, desperate men and women. Today Maleboch's descendants slumber on in their mountain stronghold, still resistant to the outside world.

On the lower eastern and southern slopes, which receive the highest rainfall, high forest occurs, with yellowwoods reaching a height of 60 metres. Old man's beard lichens hang from the long spikes of twisting liana vines, and epiphytic orchids crowd the tree branches. Since 1957 the forests have decreased by one fifth, mainly as a result of creeping fires during the dry years of the early 1960s. Recent evidence suggests the fires were started deliberately to increase the area of arable land on the flat and fertile plateaux.

The eastern cliffs shelter another breeding colony of Cape vultures, but here the interests of birds and climbers do not conflict, for the longest and hardest routes in the Transvaal – over a dozen twelve- and thirteen-pitch climbs of G standard – all occur on Blouberg's one kilometre-long north face. The first major G grade route, 'Maleboch', was successfully led in 1960 by Bob Davies, one of the strongest Transvaal climbers of the time.

In 1965 Paul Fatti, then a young Witwatersrand University student, first visited Blouberg. By opening the route 'Moonlight' he began a relationship with the mountain that still draws him back. With Californian Art MacGarr he climbed 'Moonlight Direct' in 1976 on the third attempt and in 1978 they opened 'Last Moon'. Five years later, with Jonathan Levy, he finished 'New Moon' (G1, M1), begun in 1981 with Joe Maclennan. The two most severe climbs on Blouberg must be Charles Edelstein and Alan Lambert's 'Afterglow' (G2, M-) and Eckhard Druschke and Dave Cheesmond's 'You Only Live Twice' (G3, M2). It took Druschke and Cheesmond two and a half days of sustained climbing to complete the ascent, using the siege tactics that characterize big-wall climbing. No second ascent is known of this route, which follows a straight line up the most formidable section of the cliff to a height nearly twice that of the Johannesburg TV tower.

The north-eastern spur of the Blouberg leads on to Marais' imagined desert city, from *Waar Tebes in die Stil Woestyn*. This is the Soutpansberg, named after the large salt pans on its western tip, an arid land of hollows and dusty hills. Extensive faulting of the quartz, shale and sandstone beds has produced a series of broken, parallel ridges with southerly escarpments similar to those of the other north-western Transvaal ranges.

In the Soutpansberg rainfall varies from high to extremely low. While Entabeni in the east receives some 700 millimetres of rain a year, the Brak and Sand rivers in the west tell of dry and dusty regions. The northerly slopes decline gently to the baobab and thorn savannah of the sun-beaten plains that march on into southern Zimbabwe and eastern Botswana.

The Great North Road twists through the greener crags of Wyllie's Poort north of Louis Trichardt and onwards to Beit Bridge, while in the west the road disappears along the old hunting trail known as the 'Punda-ma-tenka' road. Its name, meaning 'pick up and carry', refers to the white hunters' custom of using local tribesmen as porters for their expeditions. Here the few pans attract a multitude of birds from hundreds of kilometres around and provide a seasonal oasis on the edge of the Kalahari. Pelicans and flamingoes come from afar to nest, while ducks, coots and crakes paddle among the reeds or scratch in the water-grasses for food. Jacanas hop along the lily pads, and at sunset sandgrouse flock to the water's edge before whirring off to roost for the night.

The Soutpansberg curves gradually to the south-east and finally sinks into the sweltering Lowveld. Here the vegetation changes to mopane and acacia savannah, where rivers flow into occasional deep, green pools, still frequented by hippo and crocodile. Fish eagles shriek from their watch towers in high acacias and cormorants stand like ancient Egyptian friezes in the sun, drying their wings.

Fundudzi, the sacred and reputedly enchanted lake of the Venda, lies at the eastern end of the Soutpansberg. Few

34

outsiders are granted the privilege of visiting the lake and all who do so must pay their respects by turning their backs to it and bowing. Entire villages are said to lie beneath its murky waters, where dwell 'Ditutwane', the shades of departed ancestors and guardians of the mountains: this is the Venda people's spiritual home.

Traces of the first Venda settlements can be found near Lake Fundudzi. Of the numerous ruined canals and walled towns the most impressive is Dzata where, under Chief Ndyambeyu, these people reached the height of their power. Their influence spread far and much of the stone used in their constructions was carried there by subjected tribes paying homage to Ditutwane – the spirits of the mountains.

Here, at its apex in the Soutpansberg,

our spiralling journey turns out on itself and unfolds into another enigmatic corner of Africa. Where the north-western Transvaal is hot and harsh, the north-eastern reaches of our journey bring us into a land of waterfalls and mists, and legends thousands of years old.

34. *A rock-splitting fig plays its small part in the erosive cycle of the Pilanesberg.*
35. *Mountain streams and 'living fossil' tree ferns add a touch of magic to the kloofs of the north-western Transvaal. Few people know of the beauty and extent of these wooded, watered retreats.*   **36.** *The roots of a wild fig are strung between bole and rock like the strings of a giant harp.*   **37.** *A stream relentlessly scours a pothole pool into the hardened quartzite of Retief's Kloof.*
38. *(Overleaf) This mountain hideaway of the hereditary Maleboch chiefs on Blouberg is still cultivated by remnants of the Hananwa tribe.*

35

36

37

THE EASTERN TRANSVAAL
LAND OF MISTS AND MYSTERIES

*ld men with eyes like his are right to
  claim*
*hey know the country better than the
  palms*
*f their own hands: they are not eyes that
  seek*
he eagle's view, to read in distances
he cantos of a continent: they are
iviner's eyes that read in cryptic signs
he formula for rich discoveries:
. scale of lichen or a patch of moss,
. rocky outcrop, the minute
.ctivity of ants, a garnet chip –
*he small things, close to earth, that other
  men*
'hink trivial or do not see.

harles Eglington
ld Prospector

.old! It was the saint-seducing yellow
.etal that turned, albeit briefly, the
astern Transvaal from a remote part of
.e country into an area throbbing with
.achinery, where dusty miners fought
.esperately over claims. One of the few
.rospectors who still lives near Pilgrim's
.est, ever in search of Eldorado, says the
.round is 'rotten with gold', and indeed it
.s. But the metal has been spread far and
.hin in the rivers and soil, and the area is
.o longer considered the commercial
.ining proposition it once was.

While the young Rider Haggard was a
.eputy on the staff of Sir Theophilus
.hepstone, Administrator of the
'ransvaal during the British occupation,
.e fell in love with this mountainous
.astern corner of the Boer republic. His
.ook *King Solomon's Mines* romantically
.inked the old legends of vast gold
.eserves in the African interior with the
.eautiful Wolkberg, the 'mountain of
.louds'. John Buchan's *Prester John* was
.lso set in the Wolkberg and around the
.earby Iron Mountain.

39. *Red walls and wooded scree slopes* 40
*dominate the Mohlapitse valley below
Wolkberg.* 40. *An emperor butterfly is
attracted by the sweet-scented flowers of a
brittlewood tree.*

A northerly arm of the Strydpoortberg,
the Wolkberg is well named. For most of
the year thick cloud hangs around the
quartz and dolerite ridges, of which
Serala, at 2 128 metres, is the highest.
This is really the northern limit of the
Transvaal Drakensberg before it descends
into the dank, wooded kloofs near
Tzaneen. To the east stretches the
Lowveld, while immediately to the west,
in contrast to the forested mountainsides,
lies the hot Mohlapitse valley.

'If ever, in imaginary discourse with

one's inner wilderness, one could
conceive of a scree slope beyond the rings
of Hell, this was it,' climber Keith James
writes of Mohlapitse. 'Burned, thorned,
frustrating memories of imagined mamba
holes, malevolently hooded cobras
looming out at every bracken patch,
insidious boomslang in every face-
scratching bramble.'

That is what one begins to realize about
Mohlapitse – it gets to you on all levels.
'Let not the reader think this description
of scree is mere indulgence. It is that of
course, but much more too. For often in
Africa one cannot talk merely in terms of
rock. The scree slope of Mohlapitse and
the 100-metre krantzes above are a
veritable "tale of power".' At Mohlapitse
James and his climbing party called their
route 'Whimper': 'a humiliating
experience – an encounter, no less, in a
sacred place'.

About a hundred kilometres south of
the Soutpansberg and east of where
Pietersburg lies prone on its Bushveld
platform, Duiwelskloof and
Magoebaskloof guard the path to the
Strydpoortberg, to misty Wolkberg and
the wild Mohlapitse valley. In 1951 the
Transvaal section of the MCSA put the
finishing touches to a rough stone hut in
the Serala Wilderness Area above the New
Agatha State Forest. Hidden in an
evergreen plantation, the hut stands
beside a trout stream at the base of
Wolkberg's main crests and offers
welcome respite for mountaineers.

Long strands of old man's beard lichen
sway from aloe and tree branches on the
southern mountain slopes. Sunbirds
shimmer red, blue and metallic green as
their long, curved bills probe the krantz
aloe's scarlet flowers for their nectar. Buff-
chested Gurney's sugarbirds, also with
slender, curved bills, sit atop the velvet

blooms of *Protea rhodantha* that are common in the Wolkberg.

A strenuous but pleasant hike through the Wilderness Area leads up and over the Serala peak, along its south-eastern slope, past deep gorges with plunging waterfalls and marshy groves, down the near-vertical Kruger's Nose, over the Knuckles and finally into an area of tranquil isolation among cycad- and fern-fringed streams and mountain pools. In the dense forest that fringes the grassy knolls only screaming samango monkeys disturb the serenity.

To the east, the lush Lesitele valley descends to the oppressively hot Lowveld. The river's waters nourish the fertile lands that bear juicy mangoes and pawpaws, avocado pears, bananas and nuts. Southwards to The Downs and

beyond, a crinkled landscape extends all the way to the Abel Erasmus Pass and Olifants River valley, its spurs and valleys running at right angles to the escarpment. The tranquil beauty of the Wolkberg contrasts with the Transvaal Drakensberg's mighty drop, although the eastward-facing cliffs are still hidden from view.

Often remote and never obvious, the eastern Transvaal's climbing areas were some of the last in the country to be tackled. As late as 1959 Mitre Buttress at Umkomani yielded four or five good routes, while at the same time Ron Kinsley and Merv Prior were pioneering the Blyde Turrets (also known as the Rondavels) that tower above the Blyde River Canyon. A year later they climbed the nearby giant's staircase of Swadeni

wall, and since then some fine country routes have been explored. Recently climbing has been confined mainly to Wolkberg and the two favourite escarpment kloofs of Gibraltar and Sekororo near the mining settlement of Penge.

From the low-lying Bushveld rise foothills to about 900 metres. The true montane region, however, begins in the mist belt above 1 050 metres, where drainage is often poor and the water that collects in depressions causes swampy conditions. Characteristic water-loving vegetation grows along streams, and many endemic cycads occur in forests and thickets on rocky slopes and in gullies.

Further to the south, in a corner between the hilly area around Dullstroom and Lydenburg, the green hills of

Swaziland and the flat bush of the Lowveld, lies the Barberton Mountainland. Local names such as Kaapsehoop, Kaapmuiden and De Hoop suggest the anticipations of the weary trekkers from the Cape, coming to the end of a long journey in search of a Promised Land. This was also the old hunters' route to the coast of Portuguese East Africa, and the last respite before the drop into the fever-sodden plains that caused so much suffering for the European explorers and colonists who ventured there.

Many tales of this ancient land have been spun. The Africa of buffalo and lion, of hunters and their wagon trains, the land where Jock of the Bushveld lived and died has been vividly described by Percy Fitzpatrick. Other names, such as Revolver Creek, Joe's Luck, Eureka,

Handsup and Sheba, tell more frantic tales – of gold. Barberton itself began as a frontier mining town complete with saloons, gunslingers and nugget-bulging saddlebags.

Gold was discovered in the eastern Transvaal mountains near Sabie in the early 1870s but is was to be another ten years before the real gold rush began. That breed of man, soon followed by a similar breed of woman, that seeks adventure and fortune was drawn from all over the world along the flux of the shiny metal's mysterious magnetism. Although many gold-bearing veins and alluvial finds were to cause a rush, the main source soon dried up and the Witwatersrand became the magnetic pole of the world's gold-seekers, leaving the old hills to sleep once more.

41. *The Lisbon valley, where Billy Davis discovered the 'wall of gold' in the 1880s. Prospectors still talk with awe about Davis' legendary strike.* **42.** *Richard Jooste pans for gold in Lisbon Creek. Panning is used mainly for sampling an area's potential yield, but nowadays Richard prefers to spend his time fishing for trout.* **43.** *Peacock pyrite glints in the sun. Usually called 'fool's gold', it does contain a tiny percentage of the precious metal.* **44.** *Old-time gold mining maverick George Marshall has been mining in the Lisbon valley for more than 40 years. He has tried many methods of recovering the thinly distributed metal but is really interested only in finding nuggets. His partner, Richard Jooste, is currently working a number of claims but the two men lack the capital for a large-scale operation.*

43

42

44

Great legends and mysteries have always surrounded gold, and no less so the reefs and veins of the eastern Transvaal. This is where the mythical Queen of Sheba is supposed to have ruled over great wealth in the land of King Solomon's mines. Tales of strikes and hidden fortunes, rich beyond imagination and now lost to time, still haunt the memories of the few old prospectors who probe the caverns and rock strata. These men live as hermits in the woods, their quiet lives spiced with keen rivalry. As the years tick by the yields of the precious metal grow smaller, but the diggers' belief in the legends and tales of old burns on.

George Marshall is such a man. He lives in a ramshackle cottage in the Lisbon valley, a beautiful area between Pilgrim's Rest and Graskop. He is patiently building a new daub and wattle home, but as his partner says, 'time passes George by'. I was a teenager when I first met George Marshall and his playful Border collie and, as he had made a deep impression on me, Clive and I sought him out once more. Time has indeed passed him by – and Eldorado is still his consuming dream.

These golden hills have slumbered long, longer than anyone knows. Greenstone, Granite Hill and Boulders near Barberton are the names of places that lie on the earth's very foundation. The Nelspruit granites are part of the Primitive System of earliest rocks that have been covered by ice and by desert sands. Magma from the earth's deep forge has surged over them, and at times they have been exposed to the sky, at times smothered under thick blankets of sediment. More interesting is that the oldest known life forms have been found in the sediments lying directly on the greenstone. Blue-green algal fossils that date back nearly as far as the rock itself give an indication of evolution's patient progress.

The Barberton Mountainland is broken, hilly country, but it is the Transvaal Drakensberg to the north-west that is the real mountain country of this area. From a topographical map this escarpment appears to be a northern extension of the Natal Drakensberg but the ranges belong to entirely different ages and geological systems. While the Natal mountains are the result of geologically recent Karoo deposits being continually eroded back, their Transvaal namesake is far older and composed of different rock types.

An obvious contrast between the two areas is that in the Transvaal there is no Little Berg of outlying ridges and plateaux that characterize the Natal Drakensberg. The more northerly escarpment rises

46

45. (Previous page) *A view of the escarpment near Swadeni from the Lowveld. The Transvaal Drakensberg is formed by a band of Black Reef quartzite that has been forced up along its edge by the weight of Bushveld Igneous Complex deposits to the west.* **46, 47, 48, 49.** *Innumerable streams, waterfalls and pools are the most striking features of the eastern Transvaal mountains. Having risen in the forested kloofs of the plateau, they leap in giant steps to the hot plains below.*

directly from the Lowveld to an overall height of 1 300 metres in places. It is a raised layer of rock, sheared clean from its base and thrust upward along the line of shearing by the weight of successive layers of the Bushveld Igneous Complex that covers a large part of the Transvaal interior.

It is as if a huge finger had pushed down in the centre of a saucer of receding rock layers and the outer rings had popped up from the pressure. The outermost ring is composed of the hard Black Reef quartzites at the bottom of the Transvaal Sequence and rests on the archaean Basement granites. It was laid down probably as marine sandstone deposits in Precambrian times, about 2 000 million years ago.

The next layer of the Transvaal Sequence is dolomite, formed from carbonates precipitated by blue-green algae. The dolomite is exposed in many places in the Transvaal and, because it is susceptible to chemical action by water, can be identified by the numerous caves and sinkholes that have appeared in it. Originally on top of the quartzite, after the formation of the escarpment the dolomite was left in a band just west of the Drakensberg. The Wolkberg, Echo and

Sudwala are well-known cave systems in this dolomite band.

At one point the escarpment is cleft by a deep river gorge, so impressive that it ranks as one of the subcontinent's natural wonders. From its confluence with the Treur River at Bourke's Luck potholes, the Blyde River has carved a 16 kilometre-long canyon through the hard quartzite. The Old Mine just downriver of the potholes, where in the thunder of the gorge the river starts to race, is one of the few breeding colonies of the bald ibis.

Like so many others in the region, the names of the Treur, 'sorrow', and Blyde, 'happy', rivers have a story to tell. In 1844 Hendrik Potgieter was on his way to Delagoa Bay with a group of trekkers. At the edge of the escarpment he decided to move on more quickly with only a few horsemen and left the main body of the group camped beside a stream. When their leader failed to return, those left behind concluded that he had perished on the fever-ridden plains below and sadly broke camp, naming the stream 'Treurrivier' as they left. Some distance to the west they were overtaken by Potgieter and his men, and promptly named the river nearby 'Blyderivier' to commemorate their joy. Bourke's Luck, at

49

the confluence of the two rivers, is named after an enterprising prospector who correctly surmised that the magnificent potholes, scoured out to a depth of many metres by rocks carried by the gushing water, would be a natural prospector's pan for the heavy golden metal.

The highest point of the eastern Transvaal, and in the province as a whole, lies to the west of the escarpment's edge, among the rounded hills near Dullstroom. From around Die Berg (2 249 metres) flow innumerable streams and rivers that fall in trains of lace over the massive lip. As well as being excellent trout fishing country, this is a land of waterfalls great and small, with deep pools and rockslides that cut through forested gorges, and water plunging, lilting and splashing, cascading down.

The views from God's Window and Devil's Window on the escarpment edge reveal the sudden, dramatic drop to the Lowveld as well as vistas that stretch to the Indian Ocean 200 kilometres away. With no pinnacles and outrunning ridges, the Drakensberg here forms an impressive broken wall of rock running from south to north. Only at the opening of Blyderiviierspoort at the Swadeni wall does it briefly turn west, allowing the Blyde River to flow between the block-like Mariepskop (where chief Maripe fought off a Swazi *impi*) and the Swadeni bastions.

To the north-west of Swadeni, where

creepers and heavily spiked liana vines as thick as human limbs menacingly guard the forest's dark recesses. Among the white stinkwood, the fluted-trunked wild figs, giant matumi trees and the many other species, few of the once-prolific yellowwoods remain. Over the years their prized timber has been used to repair wagons, for furniture, and even for mine props during the gold rush.

Trees of one of the larger species, the forest mahogany, are conspicuous in that the bark has often been stripped from the lower trunk. Like that of the fig, it is widely used by Africans for medicine and magic. The sweet-scented, bell-shaped flowers are borne in the trees' dense crowns, and the wood of this large and handsome evergreen is favoured for carving.

Splashes of yellow *Ochna* flowers appear luminous in spotlighted brilliance where the sun breaks through the canopy. Bright flowers and birds inhabit this greenery but it is the butterflies that attract most attention. The forests host numerous species of butterfly which flutter in and out of sunlit pools like showers of coloured petals. Members of

e Olifants River cuts through the scarpment on its westward journey to in the Limpopo River, the Manoutsa ountain stands sentinel. On the ledges its cliffs the largest known breeding lony of Cape vultures has been onitored since 1973 by the Vulture tudy Group, which is attempting to cord the vulture's breeding pattern. An timated 600 breeding pairs and a large on-breeding population roost on the arrow ledges far above the rough ree slope.

Of the many threats to this bird's xistence, the lack of carcasses, coupled ith the disappearance of the scavengers hich crushed the bones into air-eightable pieces, is perhaps the most erious. The lack of bones has led to evere calcium deficiency in the vulture hicks' diet, with the result that their own ones are brittle and often diseased. When he young birds attempt their first flight he strong updrafts of wind, instead of arrying them upwards, sometimes snap heir wings and the fledglings plummet to itter the scree below. Vultures have been nown to swallow even shards of rockery and glass in place of the bones heir diets lack. Electrocution on high-oltage cables, drowning in reservoirs and oisoning by ill-informed farmers also ake their toll.

Monitoring the vultures is valuable work, but abseiling down and then prussicking back up hundreds of metres of swaying, bouncing rope, hour after hour in the burning sun, is hard slog. When hot and bothered field workers, precariously balanced on narrow ledges, have animal intestines regurgitated over

them by smelly chicks, they may feel like wringing necks rather than legs.

North of Manoutsa the escarpment turns again to run north-west all the way to Serala and the Wolkberg. There are few roads in this remote section of the Transvaal Drakensberg and most of the escarpment lies within The Downs Wilderness Area which is controlled by the Lebowa Forestry Department. Only the more determined hikers and climbers overcome the sheer difficulty of getting there, and to those who do so it is a jealously guarded sanctuary of high ground on a curve in the Olifants River.

Molomanye, Ga-Selati and other young rivers rise on this mountain ridge and flow in both directions to the valleys on either side. Gibraltar Kloof is a favourite yet rarely visited haunt of Transvaal rock climbers. The usual approach into the kloof is from Penge where one follows the stream up a peaceful rural valley and along fertile fields before climbing the steep side of the gorge. For the more adventurous, another approach is from Trichardtsdal to the east. Above this outpost, on the plateau above the rock towers, stands a lonely cairn of stones that is little known and seldom visited. It marks the last known outspan of Louis Trichardt before he and his by then depleted party of trekkers descended the Drakensberg *en route* to the sea, never to be seen again.

Had the restless Trichardt returned, he would have found the uncomfortably hot Lowveld yielding suddenly to fresh and fecund forests that cling to the cloud-covered bastions. Beneath the high tree canopy, lost in a glaze of filtered light,

50. *Impenetrable forest covers much of the lower escarpment and makes mountaineering there virtually impossible. Some of the country's largest indigenous forests occur in the eastern Transvaal.* **51.** *For the last decade colonies of the endangered Cape vulture have been monitored in an attempt to understand their breeding patterns.* **52.** *Mountaineers need to be on constant alert to the thieving ways of vervet monkeys, the pickpockets of the wild.*

the subfamily Charaxinae, as well as being strong fliers, are among the most brilliantly coloured varieties. Many species are endemic to the Transvaal Drakensberg, some even being restricted to a single hilltop or patch of relict forest. For example, on Mariepskop a reserve has been set aside for *Charaxis mariepsis*. These small populations are especially

susceptible to the depredations of unscrupulous commercial collectors, as are the rare and prized 'living fossil' plants, the cycads.

These trees are fascinating relics of plants that grew 150 million years ago. Members of the genus *Encephalartos*, in particular *E. longifolius* of the eastern Cape, are sometimes called 'bread trees', as Hottentots fermented the pith and used it as bread meal. The kernels, however, are highly poisonous. Like the butterflies, some species are confined to small areas, among them the Modjadji cycads in Duiwelskloof, where southern Africa's only cycad forest has been protected by generations of rain queens, known locally as 'Modjadji'. Other eastern Transvaal

species include the Barberton, Kaapsehoop and Lydenburg cycads, and there is also a species in the Waterberg which was named after its discoverer – *Encephalartos eugene-maraisii*.

The road that grinds up the valley leading to Trichardt's cairn now not only carries the forester's four-wheel-drive vehicle, but is also a major trading route the local people, who move along this track like worker ants under their heavy loads. Built by early prospectors and named 'the Balloon Road' after the farm on which it lies, it is concealed in the narrow valley until it emerges on the rocky slopes, out of sight of the surrounding country.

In a high valley behind Magakolo

53. Schizostylis coccinea, *commonly known as the kaffir lily, is an iris species that is found along mountain streams in the Transvaal and Natal.* **54.** *An embroidery of natural textures in the Serala forest reserve.* **55.** *First light of a winter's day slowly melts the frost on Mount Anderson, the highest point on the Transvaal escarpment.*

53

54

56. *Edged with riverine forest, the Mohlapitse River turns under stepped crags and around a sun-burnished spur.* **57.** *The Ga-Selati River murmurs down its gorge in The Downs Wilderness Area near Gibraltar Kloof.* **58.** *(Overleaf) Rolling hills and interlocking spurs on the Fanie Botha Trail in the still moments of late afternoon.*

mountain lies a tiny hamlet. Although remote, the place is well visited, for here lives the local 'doctor' – part *sangoma* and part apothecary to local tribesmen, who travel long distances to consult him. The place is deeply rooted in superstition, and the local people fear the power of the mountain gods: come a dark storm and the gods are angry, they say. And when lightning strikes a person it is said to be bad *muti*, a spell cast by an evil-wisher.

The spectacular heights above Gibraltar Kloof, between Trichardtsdal and Penge, are covered in mountain grasses and wild flowers. A mountain 'stepping stone', this region has floral affinities with Malaŵi's Mulanje mountain and the mighty Ruwenzori range of Kenya and Uganda. Among the specially adapted vegetation, the delicacy of the lilac *Harveya* flowers belies the plant's true nature, for it is a root parasite of coarse green erica. The southern, more shaded slopes are scattered with patches of forest where light green *Dracaena* plants sprout near streams that trickle over moss-wrapped stones, and the banks of shaded pools are sprinkled with white and pink-flushed *Gardenia* blossoms. Small birds, such as long-tailed malachite sunbirds and grassbirds with their spiky tails, flutter and flit. Raptors patrol the skies, the most conspicuous being ungainly banded harrier hawks, rock kestrels and snake eagles.

From the higher summits the land falls away in parallel ridges, dropping suddenly into the furrowed gorges that are legendary among climbers and hikers for their waterfalls and pools. In one of the three 'fingers' that branch out at the top of Gibraltar Kloof is a pool from which falls a long plume of water into the gorge below. On a sunny day it is pleasant to lie in the placid 'pool with a view' and gaze at the beautifully backlit mountain range.

In Sekororo Kloof, where a series of rock slides corkscrews down through consecutive pools, the high rock walls form a steep and inescapable ride of exhilarating speed. The 'bumslide' has in jest been graded G – the equivalent of a severe and committing rock climb. In the lower kloof wild fig tree roots bunch and twist their way over and through the rocks into the enchanting pools.

Like Tennyson's *Land of the Lotus Eaters* where 'the slender stream all along the cliff to fall and pause and fall did seem', mist-hung valleys and evergreen forests of the eastern Transvaal mountains are washed by a thousand rivers and waterfalls, of which the grand Mac-Mac Falls, the Berlin, Lisbon and Bridal Veil falls are only some of the better known and more accessible. All are too numerous to name or remember, but each adds to the rhythm and sparkle of the mountain spirit.

When early morning mists hang along the Treur River valley, when otters glide along the river, rhebuck break from their nocturnal cover or leopard spoor present a thrilling sign along a damp path, this wilderness is rich and comforting. God's Window may well be as much a secret place for him to peep in on this wonder as it is for us to look beyond.

LESOTHO AND THE NATAL DRAKENSBERG
FIRE OF THE DRAGON

These mountains of up-pointed spears
Hold eland, oribi and rhebok
Capering over yellow rock
To sandstone caves that form a barrier. . .

The painted bushman aims his bow,
The real sunset starts to flow

Across this sweeping mountain range
And still, despite ten centuries' change,
Art remains a kind of hunt
Eliminating fear and cant.

Alan Ross
Rock paintings, Drakensberg

At the time of the deposition of the Upper Beaufort Group's sedimentary layers, about 200 million years ago, the area that is now Natal was a flat, stepped landscape with scattered marshes. Across it roamed amphibious reptiles and the earliest ancestors of mammals: the short-limbed, scaly carnivore *Cynognathus*, the squat reptile *Lystrosaurus* with its tusked head, and the giant, 60 metre-long dinosaur *Massospondylus* all left their skeletons and footprints in the soggy ground near Giant's Castle, Leribe and other parts of the Drakensberg.

Geologically the Natal Drakensberg is the youngest remaining massif of the Karoo Supergroup, a 1 500 metre-thick basalt topping to the various sedimentary layers of the Little Berg. On top of the Beaufort sediments, the Elliot (formerly known as Red Bed) and Clarens (Cave) Sandstone formations indicate that drier times set in, and few fossils are to be found in these rocks. Then, at about the same time as the southern landmass known as Gondwanaland began to break up some 160 million years ago, huge flows of lava poured out of fissures in the

59. *Makhela's Kraal lies on the rich pastures of the Little Berg, below the Cathedral range.*

ground in quick succession. They probably reached as far as the present coastline of Natal and individual outpourings varied from a few centimetres to many metres thick.

After about 20 million years the flows stopped and the resulting basalt has since been eroded back at a rate of about one centimetre every five years. It is fairly resistant and forms not only the high prominences of the main range and the Lesotho highlands, but also the hard capping of the Little Berg. Once this cover has been removed, the sandstone erodes rapidly, as is seen in the steep valleys and gorges that cut into the Drakensberg. Therefore this is not a mountain range in the usual sense, but a high escarpment being subjected to downward and headward erosion.

Millions of years of erosion have pushed the high cliffs back and left many outlying pinnacles, buttresses and ridges, detached from the main escarpment but of the same spectacular height. The humbling scale of the Drakensberg is not experienced anywhere else in southern Africa: when one thinks of mountains here, it is the Drakensberg's grandeur that first comes to mind.

The main rim of the escarpment averages 3 000 metres above sea level, rising to 3 482 metres near the top of Sani Pass, where stands Thabana Ntlenyana, the highest point in Africa south of the equator. To the north-east the highest peaks on the Drakensberg watershed are Mafadi Peak behind Injasuti Buttress, at 3 459 metres, and Champagne Castle behind Cathkin Peak, at 3 374 metres above sea level.

From the watershed the Lesotho plateau dips over broken mountain country steadily down to the west, where it is bordered by the Maluti mountains.

60. *A rough stone hut shelters Basotho shepherds tending their flocks high in the Maluti range during the summer months.*
61. *In the village of Ha Ramosebo, the makers of Molianyeoe hats have been practising their craft for generations. The hats are traditionally worn by chiefs during tribal court sessions, and their name means 'one who passes sentence'.*   **62.** *An old lady of the Semokeng district grinds maize which has been traded in the lowlands for the wheat grown in her area.*   **63.** *The Senqu River, which rises in the Mnweni area, constitutes the main headwaters of the Orange River.*   **64.** *(Overleaf) The Lesotho highlands, beyond the Caledon River valley, were for many years Moshesh's mountain stronghold.*

Resting like a crown on the high tableland, it forms a natural fortress in which lives a pastoral nation in relative isolation from the rest of the world.

This nation, the Basotho, was brought together by Moshesh out of the turmoil of the *Difaqane*, the 'forced migrations', when as well as Mzilikazi other chiefs fled westward from Shaka's wrath. Among these were Matiwane, whose Amangwane tribe decimated the peaceful Mzizi clans in the Little Berg, and Mpangazita, chief of the Hlubi, who crossed the Drakensberg and invaded the land inhabited by a Sotho tribe, the Batlokwa. In turn the Batlokwa, led by the indomitable Mantatisi, plundered the area to the west.

Mantatisi, a tall, straight, lean woman, reputedly of exceptional intelligence, was utterly insensitive to human suffering and soon became one of the most feared

leaders in those violent times. Ousted from her territory, she led her army up the Caledon valley among the sandstone foothills of the Maluti range. There her hordes attacked Butha Buthe, a well-watered, flat-topped hill on which lived Moshesh, chief of a minor Sotho clan. Fearing defeat, Moshesh stealthily moved his followers to a safer home, Thaba Bosiu, the 'mountain of the night', where he built an impregnable stronghold.

The weaker tribes were continually attacked and their cattle and crops pillaged. The country was plunged into despair as slaughter and famine increased; refugees drifted aimlessly across the land in search of food and shelter. No crops grew along what had been valleys of plenty, no herds grazed peacefully in the pastures. Starvation eventually drove people to devour their slain enemies, then their fallen comrades, and later their own family members who succumbed to the ceaseless trekking. Cannibals formed themselves into hunting bands that went out raiding for fresh supplies. Moshesh's grandfather, Peete, fell prey to one such band in the vicinity of present-day Leribe. Surviving members of the once-peaceful Bafokeng, the 'mist people', became such vicious hunters that even today they are known as 'Marima', the 'cannibals'.

Most of the cannibals fled into the mountains and occupied the sandstone caves that had once been the homes of Bushmen. At Mamates the largest caves in Lesotho were once the haunt of the dreaded chief Rakotswane, while caves near Mo'hale's Hoek are still called the Cannibal Caves. Ten years after the end of

the *Difaqane* the French missionary Arbousset found abundant sheep, cattle and crops for the people along the Caledon valley, but still they were living as cannibals.

Meanwhile, Moshesh had gathered many refugees of the *Difaqane* on Thaba Bosiu and welded together a new tribe, known as the Basotho. From the security of his mountain stronghold he eventually managed to drive Matiwane, recently defeated by Mzilikazi's Matabele and thoroughly demoralized, from the Caledon area. In the winter of 1831 selected Matabele regiments travelled to the land of broken tribes. Three years of drought after the *Difaqane* had left the remaining people of the Caledon valley pathetic and destitute. Finding nothing worth plundering, Mzilikazi's *impis* moved southwards to the domain of Moshesh, now a prosperous chief. The mountain fortress he had chosen as his home proved impregnable, and two Matabele attacks were repulsed. Later,

when the Basotho came into conflict with white men over border issues, neither Boer nor British forces could dislodge Moshesh and his people.

The first known human inhabitants of Lesotho were the Bushman hunter-gatherers of whose past we know so little and yet whose passing is so deeply regretted. Cannibalism, continued tribal wars and finally the onslaught of white settlers caught the Bushmen in an ever-closing trap and they were mercilessly hunted and exterminated. This Late Stone Age culture, which has survived into the Space Age in some remote places, was incessantly victimized because its ways were not understood, its understanding of nature not appreciated and because the Bushman 'refused to be tamed'.

It is not known how long the Bushmen dwelt in the sandstone caves of the Little Berg, but for many centuries this was a paradise where they lived in harmony with bird and animal, snake, flower and stone. They grew no crops and domesticated no animals, yet lived among the plants and beasts with an intimate knowledge of all they saw. They inoculated themselves against snake bite and knew every poison and delicacy in their environment. Modern science is often at a loss to explain what the Bushmen took for granted. Long before European culture knew about the moons of Jupiter, the Bushmen told stories of them: the stars were the campfires of departed souls that wandered across the heavens, forever hunters of the skies. They told of when the land had been flat marshland – as we now know it was millions of years before their time, in the days of dinosaurs.

With all their knowledge these diminutive hunter-gatherers had the simplicity and cheerful disposition of children, were generous souls who denied material possessions and upset nothing in the ecological balance of their surroundings. Most interesting, though, was their love of dancing, story-telling and, of course, painting. The last Bushman known to have been shot in the Drakensberg, in 1866, was one of their artists: around his waist was found a leather belt on which hung ten antelope horns containing the various pigments used to adorn cave walls. The Bushman paintings found in southern Africa exceed all other cave paintings in the world in both quality and quantity – and nowhere more so than in the Drakensberg. An artist who studied and loved their work, Professor Walter Battiss, said: 'No artist has said more, saying less.'
Although today we marvel at the way in

65

which the Bushmen bridged the gap between human reasoning and the instinctive behaviour of animals, the chauvinism of the white colonists regarded these children of the earth as savages, wild and hardly human. The ploughs and guns, the herds and horses of the white invaders tore up the Bushmen's Eden. By 1890 there were no known Bushman survivors in the Drakensberg or Lesotho, although years later signs of their presence were still occasionally found. In 1903 the Giant's Castle Game Reserve was proclaimed, and one wonders how, if they had survived for a few more decades, the Bushmen would have fitted into the nature reserves of today, for their wise use of the land and its resources seemed instinctive. They apparently destroyed nothing but for their survival and there is evidence to suggest they practised controlled veld-burning in the Drakensberg to rotate the feeding patterns of the wild animals, an example of 'agriculture' advanced for Stone Age man. It has been observed how well they tended the delicate Little Berg, while invasions by both black and white farmers soon led to overgrazing, ploughing, tree-felling and excessive veld-burning.

The Little Berg cannot take much abuse. Even the paths made by hikers in the more popular trail areas are taking a heavy toll of the thin and slippery ground cover. In an attempt to conserve the natural resources of the Drakensberg the Natal Regional Planning Commission has proposed that the area be divided into

four land-use zones. Unfortunately protection of these zones is not enforced by legislation and they are still open to misuse.

The first zone, the Wilderness Heart, extends from the top of the Little Berg to the watershed and, being ecologically fragile, should be managed primarily for water conservation. The slopes and valleys of the Little Berg make up the Landslide zone, the most fragile of all. Below this, the Trail zone has great scenic and ecological diversity and is suitable for hiking and horse riding on constructed paths. Although the zoning allows for only rustic accommodation, plans have been mooted to develop luxury resorts in this area. Finally, the Threshold zone allows more intensive land use in the form of agriculture and the provision of holiday accommodation.

The vegetation of the main Drakensberg range, between the Amphitheatre in the north and Giant's Castle in the south, is determined mainly by altitude and aspect (orientation to the sun). At higher altitudes the range of temperature extremes increases and the vegetation becomes shorter and hardier. Likewise, north- and east-facing slopes receive more sunshine than south- and west-facing ones, and this too influences plant development. Forests are more prevalent on the cooler slopes and in the damp, shady gorges, while protea savannah occurs at the same altitude but on the slopes receiving more sunshine.

The escarpment here has been categorized into three main vegetation belts: montane, covering the slopes of the Little Berg and its associated river valleys; sub-Alpine, from the top plateaux of the Little Berg to the bottom of the High Berg cliff faces at about 2 500 metres; and Alpine, including the main escarpment cliffs and the summit plateau.

Botanically the montane region is the most diverse. The profusion of species in the indigenous forests reads like a shopping list of the Great Gardener: apart from the predominant real and Henkel's yellowwoods, common trees of the

65. *A lanner falcon takes to the air near Giant's Castle. Lanners usually kill smaller birds on the wing, swooping down on them at speeds of up to 100 kilometres per hour.*
66. *An adult and a juvenile lammergeier glide past the mountains in search of food. The birds' thick layer of down, insulated by a coat of lanceolate feathers, helps conserve their body temperature during long winter nights on the snow-bound mountain tops.* **67.** *The bald ibis, most handsome of the ibises. The range of this rare bird has decreased substantially over the years and it is now an endangered species.*

mountain forests are Cape chestnut, wild peach, mountain saffron, assegai tree and Cape beech, as well as others whose names speak of their diversity and beauty.

Fire has had considerable influence on the Drakensberg's vegetation and the larger, exposed trees are most vulnerable. Grasses are better equipped to survive the ravages of veld fires as their growing points are at ground level. Fire, therefore, has tended to maintain the extensive grasslands while checking the advance of woody plants and besieging the trees and tall bushes in protected areas. In the sub-Alpine belt, which is regularly subjected to fire, certain woody plants such as the mountain cedar and the erica-like *Philippia evansii* have managed to re-establish themselves only after a period of more than 20 years undisturbed by fire. In previous times they and the forests probably covered a far greater area than they do today.

The most common species of the protea savannah are *Protea multibracteata* and *P. roupelliae*. Higher up on the more exposed slopes grows the gnarled *Protea dracomontana*, its long underground rootstock making it the hardiest of a generally fire-resistant group. The exposed slopes of the sub-Alpine region are covered mostly with the reddish *Themeda* and *Festuca* grasses, while sites protected from fire support the leggy, multi-stemmed, hard-leafed vegetation known as sclerophyllous thicket, or in the Cape, 'fynbos'. The berg cycad also occurs in this area.

In the harsh conditions of the summit's Alpine zone the vegetation is sparse and only plants with few leaves and a height of less than half a metre survive. Among the grass tussocks grow soft white, yellow, pink and purple everlastings, while in wetter areas a few flowering plants such as *Kniphofia* and *Moraea* species seasonally add splashes of colour.

The limited resources of the higher zones support few large animals other than the hardy feral horses and mountain rhebuck. Ice rats, named from their habit of sunning themselves while there is snow on the ground, are affectionately known as 'wurzles' by the mountaineers whose food packs they raid. Unmalicious theft is an accepted way of life in the High

Berg; it is a case of survival of the smartest in the chain of redistribution from mountaineer to tribesman, to baboon, raven and rat.

A few breeding colonies of the rare bald ibis, most handsome of all ibises, are found in the Drakensberg. They nest on cliff faces, usually in wooded recesses near waterfalls, but prefer to feed in the mountain grasslands. Progressively drier

conditions and overgrazing of their habitat have led to their steady decline. One colony found in the northern Transvaal suggests their range was once far greater, before a dry cycle set in about 1 800 years ago.

Some 250 bird species have been recorded in the greater Drakensberg area – about one quarter of the subcontinent's species – but few of these ever reach the

68. *From the Sentinel, the Amphitheatre sweeps down to the Royal Natal National Park. The Eastern Buttress, Devil's Tooth and Inner Tower stand above the clouds.*
69. *Below the basalt turret of Giant's Castle, the Mountain Club hut nestles beside a tarn.* **70.** (Overleaf) *Heavy August snow around Cathkin Peak and Champagne Castle; this is the Drakensberg at its alpine best.*

gh Berg. Some of the smaller chats,
rushes, swifts, sunbirds and doves are
en at this altitude but it is the birds of
ey which really belong here, and
uthern Africa has the world's largest
are of these royal birds. Black eagles,
ape vultures and white-necked ravens
habit the High Berg, but of all none is
ore magnificent than the lammergeier,
bearded vulture.

The lammergeier's range is from
southern Africa to Spain, and across
Europe to central Asia. However, it is
confined to the higher mountain areas
such as the Himalayas and the Ethiopian
highlands, the Drakensberg and the
Malutis, where it soars at great height
and speed between the peaks and
along the crags.

In southern Africa these vulture-like
eagles once inhabited the western Cape
mountains, but since 1940 they have been
confined to the Lesotho highlands, where
primitive pastoral conditions do not
encroach on their habitat. Well suited to
the cold environment, they have stiff,
overlapping feathers that act as an
effective windbreaker and cover a downy
insulating layer, 'leggings', and loose
lanceolate feathers that form a mane
on the neck.

Because of their insulating layer of
down the birds lose little heat overnight,
and therefore expend less energy
rewarming in the mornings than would be
expected of birds of their size. However,
when winter temperatures are extremely
low and all available warmth must be
utilized, fully grown birds with
wingspans of up to three metres have been
observed at the entrance of their cave
nests at dawn, their wings outstretched,
waiting for the warming sun.

By the time they reach adulthood, at
five to six years old, lammergeiers have
changed colour from general slate-grey to
a most majestic coloration: golden neck,
chest and underparts, white head, and
black back and wing tips. The red
sclerotic ring around the yellow eye is
enveloped by a black 'bandit's mask'
which extends below the chin. Here it
forms the black beard of spiky feathers
from which the English name, 'bearded
vulture', derives. The other common
name, 'lammergeier', suggests that the
birds carry off young lambs, but is
actually a misnomer, as marrow is their
preferred food. The bird drops bones,
bomb-like, from a height to shatter on flat
slabs of rock and then extracts the marrow
with its long, scooped tongue.

The lammergeier has a long, diamond-
shaped tail and in flight looks like a giant
falcon. Its numbers in southern Africa
have been estimated at as low as 20 pairs,
while more thorough, though still
incomplete research counts 300 pairs
– still a pitifully small population. In
recent years there have been few
lammergeier sightings in the Drakensberg
and then mostly by mountaineers on keen
lookout. What reward, to scale a massive
peak and to sit, exhausted and exalted,
watching the sunset over the highlands
and then to see one of these great golden
birds swooping by, like a herald from the
mountain gods. It is experiences like these
that draw a certain kind of person to the
mountains, and the most adventurous are
those who first confront and scale the
high peaks.

The watershed of the Drakensberg
forms the border between Lesotho and
South Africa, an area too inhospitable for

anyone but the transient mountaineer and blanketed shepherds tending goats and cows along the narrow, rocky pastures. Feral horses amble across the broken landscape and huddle together when the freezing winter winds rip over the Berg.

It takes a determined mountaineer to set off on a hike across the Lesotho plateau, but many are the rewards for venturing into the most rugged places. The story of four such hikers is an interesting one: 'As we drove out of Maseru the rain which was to accompany us throughout the trip began to fall. Before long we were dropped in the muddy remains of the track; our driver refused to continue. We started walking, sloshing through puddles as the rain poured down, only occasionally catching a glimpse of the mountains in the surrounding gloom. . .

'We were up early and witness to a splendid sunrise over Thabana Li Mele. The crisp air with its promise of wild, mountainous terrain to come, smoke rising from the thatch of nearby villages, and the shouted greetings of the locals across the valley made an unforgettable moment. . .

'From the river we ascended into the foothills of the Thaba Putsoa range, following the old track to Semonkeng. After a cold, windy lunch on the ridge, Jim (Benson) and Peter (Faugust) went ahead while we followed on. In doing so we provided a village with much amusement, for Jim, in crossing a particularly boggy stream, got so embedded in the mire that Peter had to drag him out, while the villagers on the hill fell about laughing. Scarcely had Jim and Peter disappeared over the hill, when we approached the same spot and repeated the performance. . .

'The Maletsunyane Falls were magnificent. The pleasant stroll to get there led through golden wheat fields, forming a dramatic contrast to the deep green, blue and purple of the gorge and distant mountains. The roar of water heralded the falls, then suddenly we were able to look down into the awesome depths of the river raging below. A light aeroplane joined the birds sweeping over the gorge. . .

'Camp that night was on a barren plateau with drinking water from a small spring in a boggy, hoof-churned field. Next morning we were woken by a horseman galloping up, his arms raised high in greeting and his scarlet blanket flying in the wind. He was the local headman, come to enquire after our health and our plans; a magnificent figure with his fiery horse snorting and prancing beneath him. . .'

71

72

Mountains have many moods which can change significantly in moments. Winter shows the Drakensberg's finest face when snow blankets its surfaces and powders its slopes. Ericas burst their living greenery through the powder, and icicles hang from cave and rock lips. The chill pierces deep into all living things, but the freshness is thrilling to the well-prepared visitor as the frozen ground crunches and crackles underfoot. This is when avid mountaineers pack their bags and head for the hills.

During the glorious summer months it is a stirring experience to sit on the edge of the escarpment enjoying the panorama below, and to watch one of the frequent angry storms that begin in the valleys and move up the slopes of the range. In hot summer weather an onshore wind from the warm Mozambique Current bears heavy, moist air inland. When it encounters the escarpment and is forced to rise, the weather changes suddenly and wild, billowing clouds shroud the peaks, breaking in furious dark waves over the cliffs. Whips of lightning that crack into the basalt spires are enough to make a hardened sinner repent, and it is not unusual to hear the resounding crack as boulders and overhangs give way and plummet to the valley below.

No-one should venture into the High Berg without sufficient food, warm and waterproof clothing, bedding and preferably a lightweight tent. Even experienced mountaineers have perished here through miscalculations or misfortune. Only 70 years ago it was believed that to be benighted on top of the Drakensberg would mean certain death. The mountains may be kind and beautiful to those who abide by their demands, but are cruel and relentless to those who flaunt them.

In true alpine tradition each section of the MCSA has organized rescue teams and the Natal section of the club is frequently called out to aid or locate people stranded, injured or killed in the Berg. In fact the club's objectives are more than just getting to the top of mountains: they include organizing and facilitating mountain expeditions; providing for the safety of mountain climbers and organizing search and rescue parties; and protecting and preserving the natural beauty of the mountains and their water supplies.

Whether telling of pioneering hikes or hazardous rock climbs, the stories of the mountaineers are part of the legend, part of the spirit of the mountains – and none more so than those of the Drakensberg. Father A.D. Kelly, one of the pioneering mountaineers of the Amphitheatre, said soon after conquering its Outer Tower (now known as the Eastern Buttress) in 1914: 'He who does not know the joy of mountaineering does not know what joy is, and this reaches its acme when it is crowned by the achievement of a first ascent.'

The warm hearth of those early days was the Rydal Mount Hotel, owned by Father Kelly's climbing partner, Tom

Casement. It slowly became run-down and by the 1960s R.O. Pearse, author of *Barrier of Spears*, recalled it as a sad and lonely reminder of the glorious days of 'whiskered men' and the 'soft swish of skirts' in the midst of a wild, unknown range.

New generations will continue to find new and more challenging routes to the top of the mountains. Because of their height and grandeur, or the epic nature of their conquest, however, particular climbs in the Berg stand out. But the mountains are never really conquered. Rather they are confronted and for a brief instant stood upon. They allow the adventurer a brief glimpse of their essence, but the favour is quickly withdrawn, to be coaxed out again by others.

The first modern exploration of the Drakensberg was undertaken in 1836 by two missionaries of the Paris Mission Society who journeyed across 'the Roof of Africa' in search of the headwaters of the Orange River. Thomas Arbousset and François Daumas travelled by pony over the Maluti mountains to the 'blauwe bergen' with their blue-grey basalts, as far as Mont-aux-Sources and the grand Amphitheatre panorama. They took licence to name the highest point there 'Mont-aux-Sources', believing it was the spring of the Orange, Caledon and Tugela rivers.

It was a mounted police patrol sent from Giant's Castle to Mont-aux-Sources in 1908 to demarcate the border between Natal and Basutoland that established the true source of the Orange River: it rises in the Mnweni area near where the gothic, fluted Rockeries peaks pierce the sky. This patrol negotiated the first known traverse of the main escarpment – a distance of about 120 kilometres. Now popular among hikers, the traverse is certainly an epic and requires at least ten days of strenuous hiking to reach Sani Pass, about 40 kilometres south of Giant's Castle, from the Royal Natal National Park.

The patrol began its journey up the Langalibalele Pass, used in 1873 by that

71. *Whatever the weather, Cleft Peak is one of the finest vantage points in the range. The view southward along the escarpment shows Cathkin Peak, Monk's Cowl and Champagne Castle in the distance.* **72.** *Climbers abseil off the first Ifidi tower after climbing an F3 route.* **73.** *Mnweni Pinnacles offer a spectacular climbing challenge. Umkulunkulu, on the extreme right, was first climbed in 1972 and Umkulu, to the right of centre, a year later. The Outer Pinnacle, seen through cloud, is one of the most difficult climbs in the Berg.*

rebellious chief in his brief war with the Natal authorities. Langalibalele stubbornly resisted interference in his affairs by either Zulu or British, and not only encouraged his tribesmen to secure firearms, but also planned open revolt against British rule in Natal. When the British mobilized a force of volunteers against him, Langalibalele retreated with his men up the Bushman's River. A short but bloody battle was fought high in the Drakensberg, and the intransigent chief fled into Basutoland where he was caught, tried and banished to Robben Island. After

the first ascent of Champagne Castle, at 3 374 metres the second highest peak in South Africa, and followed this with more serious climbs, successfully challenging both North and South Sterkhorn. The Stocker brothers and Andrew Gray even did some ice and snow climbing on the stout Cathkin Peak whose African name, Mdedelolo, means 'make way for the bully'.

It was not until around 1910, when road and rail reached as far as Bergville, and the Giant's Castle Game Reserve and Royal Natal National Park had been

showing a path through the many rock banks, scree slopes and cliffs that promised 'a fine bit of rock climbing with firm rock and good holds' but with a deep abyss on both sides.

'Suddenly the slope eased off,' says Wybergh. 'We looked in vain for further cliffs above us and then realized the peak was conquered.' From then on the peaks of the Dragon Mountains fell steadily one by one as the country's finest cragsmen turned their eager eyes to the ultimate peaks of southern Africa. Other fine routes have been forged up the Sentinel,

74

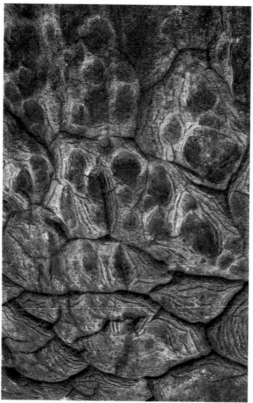

75

the skirmish the Bushman's River Pass was renamed Langalibalele Pass.

It is an indication of the Drakensberg's formidable character that less than 80 years ago the range was hardly known to anyone but the shades of the dead Bushmen whose hunting fires blaze in the heavens. They must look down with nostalgia to see those who mercilessly stole their paradise now trying to recapture for themselves a little of what was once there.

The first known climbers in the Berg were the Stocker brothers. Members of the British Alpine Club, they made numerous ascents of up to E grade (hard rock work) in the Champagne Castle and Mnweni areas. Although most of their ascents were no more than strenuous hikes up gulleys, E grade climbing in the 1880s was phenomenal at a time when only the easiest routes up Table Mountain were being attempted. In April 1888 they made

established, that climbers began their onslaught on the range. Mostly it was exploratory scrambling but not infrequently these parties encountered severe rock climbing obstacles. This was mountaineering at its finest – facing the mighty unknown with only the barest of equipment.

The first of the great free-standing peaks to be tackled was the Sentinel, a huge pillar that forms the north-eastern block of the Amphitheatre. W.J. Wybergh spoke of 'a great peak that dominates everything else and invariably draws and concentrates upon itself the longing gaze of every true mountaineer'. He carefully studied the Sentinel for a route up its steep south-west face and later returned with Lieutenant N. MacCleod of the Harrismith Garrison. On 29 September 1910 they braved a thick mist which enveloped both them and the peak. The earlier reconnaissance proved fruitful,

74. *Looper caterpillars feed on a cycad leaf in the Little Berg.* **75.** *Unusual sandstone textures in lower Mnweni.* **76.** *Ice-encrusted alpine flowers on the slopes of Mont-aux-Sources are quite used to a winter blanket of snow that is sometimes metres thick.* **77.** *Tunnel Gorge has been carved out of the soft sandstone by the pounding of the Tugela River where it hits the Amphitheatre floor after hurtling over the escarpment.*

most notably the 'South-east Arête' in 1969 by Tony Dick and Roger Fuggle. A year later Fuggle and C. Shuttleworth climbed the long and obvious crack on the North Face, straight up to the highest point on the Sentinel.

January 29, 1938 marks one of the most tragic events in the long saga of Drakensberg mountaineering. On that day Dick Barry fell to his death while retreating with his companion, Colin Gebhardt, from the awkward North-west Ridge of the Monk's Cowl.

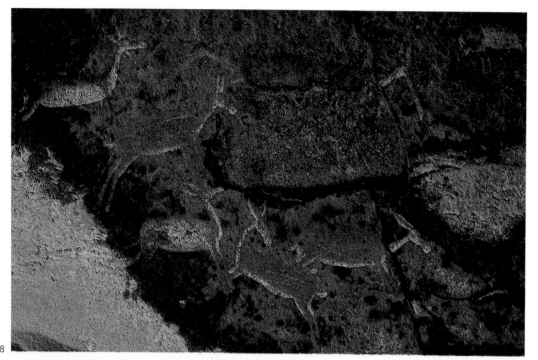

78

79

80

In 1921 George Londt, one of the strongest and most determined of the Cape's early climbers, had claimed that the Monk's Cowl's smooth walls would never be climbed – a challenge he refused to submit to when issued by anyone else. As any new generation is wont, Barry thought otherwise.

One momentary lapse of precision on the treacherous basalt of the crux pitch and 'Gebby' peeled, pulling the unprepared Barry down with him for a hundred metres. Miraculously they survived the fall but Barry was knocked unconscious. When he regained consciousness they decided to retreat down an easier route and so moved on unroped. They moved apart and off to the right, frequently calling to each other. Then there was a long silence – Gebhardt knew Barry was dead. It was his first and last trip to the Drakensberg.

Dick Barry was born in Johannesburg in

1916, schooled in Grahamstown and in 1933 departed for England to study mining engineering. From his introduction to climbing in North Wales in 1934 Barry was to prove a brilliant climber. His progress was astonishing; within two years he was leading and opening the highest grade climbs on rock and ice in England, Wales, France and the Alps. They called him 'the Tiger' – a title which is still bestowed on the élite of young climbers. In November 1936 he set sail for South Africa and on arrival began opening fine routes around the Transvaal that even today only the 'young tigers' dare follow, including 'Coffin' in Tonquani Kloof and 'Zimbabwe Tower' at Hanglip. A year later saw him slogging up the Cathkin gulley to the Monk's Cowl.

Although the peak was conquered in 1942 by Transvaaler Hans Wong and three others who found a route up the southern face, it took 24 years and four attempts before the North-west Ridge was finally climbed by Matt Makowski, Malcolm Moor and Martin Winter. Clive Ward well knows the dark treachery of that peak. He called it 'a dog's day on Monk's Cowl' when in 1978 he and Eckhard Druschke made their challenge.

'The mountain mist swirled around our peak – earlier thunder had rolled around the brooding precipices of Cathkin. From a distance Barry's route did not look good. The lower slabs were running in water so we started by climbing the much harder but drier rock to the left. The going was awkward, but we had to keep off the easier but dangerously wet rock to the right. We decided that without wetsuits and aqualungs Barry's route was all but impossible. The ridge to the left looked interesting. It was hard and the top portions most likely required some aid climbing, but most important, it was bone dry.

'Eckhard led off for about 25 metres, found a small belay and brought me up. Above us was a steep wall ending in overhangs. A break in the middle bulged moderately but looked feasible so up I went. I came to an old ring piton and shouted my surprise to Eckhard. There was a rotten hemp abseil cord attached to

Cave paintings in Knuffel's Cave (78), Ndedema Gorge (79) and Game Pass (80). The Bushman artists used powdered clays, ochres and charcoal for the predominant colours and mixed them with animal fat, egg white and euphorbia juice to make paint that, in places, has lasted for centuries. **81.** *Ndedema Gorge was one of the last and loveliest sanctuaries for the Bushmen who lived in the Drakensberg, before they were hunted out by black and white invaders.* **82.** *(Overleaf) Seen from Ndeldema Dome's Summit Cave, snow on Cleft Peak, Cathedral Peak, the Bell, Inner and Outer Horns, the Eastern Buttress and Inner Tower glows in the morning sun.*

81

– obviously the highest point some
her party had gained many years ago
[Des Watkins and Gillian Bettle in 1952].
The piton was still good so I clipped in a
runner and continued.

'The wall was now becoming serious so
arranged my two *étriers* [short rope
ladders used in aided climbing] in
readiness. I worked my way up to the
bulge and placed a piton in a crack to my
right.' Clive found a large flake shaped
like an ear and a diagonal crack above the
bulge. Using the flake as a layback
handhold and fixing his *étriers* into the
crack, he began moving upward.

'Suddenly I was descending fast.
Running belays were flashing past and
bounced to a halt, hanging free from the
rock.' He swung himself back onto the
rock face and climbed back up to where
his one *étrier* still flapped in the breeze.

Thunder began to echo around the
gloomy walls – it was time to make haste.
As Clive put his weight for a second time
on the *étrier* clamped behind the flake,
history repeated itself. 'Again I was
falling, but now I was holding half a ton of
rock across my chest. The flake in the
shape of an ear had taken on the shape of a
tombstone. As the tombstone and I parted
company it took away a piece of flesh
from my shin. The rope checked my fall
but the tombstone accelerated directly
towards a white helmet which was
Eckhard's head. At what I anticipated to
be the moment of impact the boulder
suddenly flipped into the rock wall and
shattered into four pieces, all but one of
which missed Eckhard and that badly
gashed his leg.

'Fortunately no bones were broken in
Eckhard's leg but he was in pain and
could hardly stand. We rigged an abseil
and I went over first but as
I touched ground a hail squall hit us.
Eckhard slowly hopped down the face to
join me.' What had been a simple
scramble up the grass and rock slopes, on
the return journey became a dangerous
and pitifully slow series of abseils in the
torrent. Flashes of lightning slapped into
the rock above, and the pounding thunder
nearly flattened them. After what seemed

a very long time they finally reached their
tent. By now Eckhard's leg was a little
better and he could bear his own weight
but Clive had to carry both heavy packs.
No hike in the Berg is easy and the
descent after a long climb can be torturous
even if you are in good form.

'When we were about a kilometre from
our car I was forced to stop every hundred
metres and sink down onto my hands and
knees to stop my back and shoulder
muscles going into spasm. My problem
was a temporary one and I soon

84

recovered; Eckhard ended up in hospital
with a badly infected leg and three weeks
off work.'

There have been many other epic
climbs in the Berg and no doubt more will
follow. The towering Mponjwana (the
little horn on a heifer's head) pinnacle,
also known as Rockeries Tower, for many
years defied the country's best
mountaineers. Guarded on all sides by
deep abysses, it is one of the climbing
prizes of the Berg, and was first won in
April 1946 by the individualist cragsman
George Thomson and Cambridge climber
Ken Snelson. In 1969 Tony Dick and
Rusty Rowsell put a route up this great
tower of rock that ranks with the finest
in the Berg.

The Rockeries themselves, just east of
Mponjwana, offer climbers the cleanest
rock and hardest routes in the Berg, and
are therefore favoured. The eight
pinnacles have attracted mountaineers
such as George Thomson in the 1950s,
and Carl Fatti and Tony Dick who put a
delicate route up Pinnacle G in 1969. In
the nearby Mnweni Cutback the Inner and
Outer pinnacles also provided a challenge
to climbers. The Outer Pinnacle was the
first to be climbed, by George Thomson
and Charles Gloster in December 1948,
and its companion was conquered
seven months later, by a party led by
Jannie Graaff.

Probably the most talked-about single
achievement in Drakensberg climbing is
Thomson's December 1945 solo ascent of
the Column, situated next to the Pyramid
in front of Cleft Peak. He started the climb
with a group of enthusiastic hikers but left
them at the base of the tower when he
decided to make a go for the summit. The
ascent itself was spectacular, but the
descent even more so. Having reached the
top, he found the F grade upper sections
could not be reversed without a rope. The
only projection for about 250 metres was a

85

small erica bush ten metres down. He
jumped, the bush held, and within a few
hours he was back at the base of the most
exposed pinnacle in South Africa.

Devil's Tooth, a tall rock needle
balancing between the Inner and Outer
buttresses of the Amphitheatre, was the
first spear of extremely severe standard to
be climbed. Numerous attempts over
many years failed dismally to cap it. For
long periods it was left unpicked but any
ascent that has refused some climbers
beckons to new, more daring ones until it
finally succumbs.

During the 1940s it was said that the
Tooth remained for another generation of
climbers – one that would have to be
dental mechanics to achieve any success.
The post-war period produced the first
such generation. When in August 1950 a
combined Natal and Transvaal team
comprising Ted Scholes, David Bell and
Peter Campbell took six hours to ascend
only fifty metres, they too feared they
would be defeated by the Tooth. To their
amazement the next hundred metres
proved far easier and within a few more
hours they had overcome some of the
most difficult E, F and G climbing yet
attempted and were standing on the peak.

In more recent years Carl and Paul Fatti
have climbed high-grade routes in the
Drakensberg, including the second ascent
of the north-west spur of the Western

83. *Quiet calm pervades lush indigenous
forest in the Tugela valley, Royal Natal
National Park. The predominant yellowwood
trees grow only in damp ravines of the park or
on shady south-facing slopes.* **84.** *An infant
dassie finds security on the back of its mother.
Dassies are the favourite food of many
predators and must be ever watchful.*
85. *Although a bite from a berg adder
generally is not fatal, it may cause weeks of
illness and complications such as temporary
blindness.*

86. *A view across the majestic sweep of the Amphitheatre, from the mountain homeland of Qwa-Qwa. The massive tower on the right is the Sentinel.* **87.** *Cathkin Peak, Monk's Cowl and the escarpment wall loom ominously above a bank of cloud onto which is projected a Brocken Spectre of the photographer. Considered by many mountaineers to be a good omen, this rainbow of light is an alpine phenomenon caused by refracted light cast around a person's shadow on cloud.* **88.** *From Twins Cave at the hinge of the Cathedral range, looking northwards, the Saddle, Mnweni and the eastern corner of the Amphitheatre recede like cardboard cut-outs in the twilight.* **89.** *(Overleaf) From Mbundini Buttress on the left across Ifidi to the Eastern Buttress, the escarpment awaits an approaching storm.*

Injasuti Triplet, next to the beautiful Red Wall waterfall on the Injasuti River. While Paul was at the University of the Witwatersrand, Carl bolstered the young Natal University climbing contingent. In 1968 he climbed the Frontal Buttress of the North Saddle with Tony Dick, Roger Fuggle and Barry Manicom. Paul Fatti is unquestionably the most experienced mountaineer living in this country, having led a number of expeditions to Patagonia, Baffin Island and the Himalayas. In 1982 I was fortunate to accompany his team as a photo-journalist on a first ascent in the Jugal Himal in Nepal. In 1983 Merv Prior, a Transvaal 'tiger' of the late 1950s and also a member of the Jugal expedition, opened a new frontal route up Giant's Castle.

When going up one seldom considers coming down, but Clive Ward will long remember another epic descent he had to undertake, this time while collecting photographic material for this book. He and a climbing companion, Tony Maddison, were ambling along the remote slopes of Mnweni in February 1984 when Tony was bitten by a berg adder. Within an hour his sight was fading, he was losing control of his muscles and he could hold down neither food nor liquids. In a race against time and the closing weather, a bivouac had to be found on the steep slopes and preparations were made to leave him there while Clive went for help.

The next day dawned cold and misty and the Air Force rescue helicopter from Durban searched for the little red tent,

86

ping for a break in the cloud. Luckily
e cloud did break just enough for the
licopter to make a one-wheel
uchdown so that Clive could climb out
d then guide Tony back in. A week in
spital and Tony was back on his feet,
esight fully recovered and a mountain
scue had been successfully undertaken.
Everyone who knows the Berg has a
vourite and perhaps special place along
is glorious escarpment wall. Perhaps
e isolated and remote pinnacles of
nweni and Mponjwana; the green
rests, bubbling streams and falls and
eat sweep of the Amphitheatre; or
aybe the long green valleys, Bushman
ves and dominating bastions of the
int Injasuti triplets, Ndedema and
ant's Castle.

Certainly the view from the Eastern
Buttress of the Amphitheatre is
unequalled anywhere in South Africa,
perhaps even in the world. To the north a
wide sweep of dark basalt falls a thousand
metres to the green forests and red-gold,
grass-covered shelves of the Little Berg.
The Tugela River plunges over the
escarpment and meanders across the
emerald-green plains of the Natal
Midlands, framed on the left by the
massive Sentinel.

Directly below the Sentinel Devil's
Tooth slices up, perhaps through
billowing cloud, while the rocky plateau
behind glides gently up to 3 282 metres at
Mont-aux-Sources. Looking south, a
hundred kilometres of ridges and ranges,
like cardboard cut-outs, fade into the

distance: Ifidi; the Mnweni pinnacles and
Mponjwana set in front of the sloping
Saddle; the long Cathedral range with Bell
and Horns; Pyramid, Column and Cleft
Peak; Sterkhorn, the bully Cathkin and
ominous Monk's Cowl. The massive
Champagne Castle obscures Giant's Castle
from view and still further is Sani Pass,
Sehlabathebe and finally Bushman's Nek.

Here the waterfall plunging over Ben
Macdhui is the last trickle down the
dragon's tail. There are rumours of
fissures opening recently somewhere on
the Lesotho plateau, from which hot
vapour is said to rise. The dragon
slumbers on, its old scales slowly
dropping off and its fire smouldering
beneath the skin. But it has slept for aeons
and is yet in its prime.

87

88

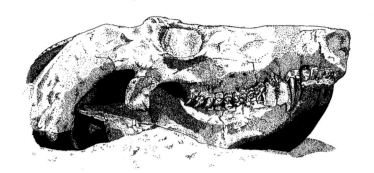

THE CAPE INTERIOR
WHERE DRAGONS SLUMBER

The rivers of the Eastern Cape lullaby
  ancient beasts:
  rubberflesh reptile, soupblood bird,
  leathery mollusc and mouse,
shrug inside black blankets of slate, and
  shrinkle slowly to stone:
fossil frosts, frizzles, splits and
  mammoths strut the human stream.

Chris Z. Mann
Rivers

It is difficult to establish exactly where the Drakensberg massif ends for its basalt forms the icing on the layered cake that is the Karoo Supergroup of rocks. Where ridged fingers fan out to the south into the mountains and valleys around Rhodes village and Barkly East, the Stormberg rises as a flat-topped range, running in a single line from east to west between Dordrecht and Molteno. At its eastern end the range bunches back up against the Witteberg, where stands Ben Macdhui, at 3 300 metres the highest mountain in the Cape Province.

South-west of Molteno the Bamboesberg curves in an 'S' deep into the sheep-farming country around Cradock, and as the land gets lower, it becomes drier, hotter and barer. Following the mountains, we begin to descend through the layers of rock that make up the Karoo Supergroup. Observant travellers will realize, as they pass through cuttings showing pastel-coloured bands of rock, they are seeing the waste deposits of the ebb and flow of time's tides.

Far to the south-east of Ben Macdhui and remote from the other mountain ranges in the province, stand the Winterberg range and, merging into them, the Amatola, whose Xhosa name means the 'place of many calves'. Instead of flat tablelands and koppies surrounded by arid veld in which aloes and sweet-thorn predominate, the Amatolas are skirted by

90. *The tail of the Drakensberg peters out between Lundean's Nek and Moshesh's Ford in the north-eastern Cape. The caves and overhangs in the river valleys of the area are rich in Bushman paintings.* **91.** *As the great hunters and guardians of the first spirit of southern Africa, the Bushmen had a communion with the mountains and their natural environment that was highly efficient yet deeply spiritual. It is said that they never killed an eland – the greatest quarry of all – without commemorating the occasion with a ritual dance.*

yellowwood forests and golden grasslands, and for a few months every winter are frosted with snow.

This 'place of many calves' was the scene of an important chapter in the country's recent history. The years immediately prior to the arrival of the 1820 Settlers in the eastern Cape saw turbulent times in that part of the land.

After numerous confrontations between trek farmers and Xhosa herdsmen advancing along this wild frontier, the British Army built a line of forts to monitor the border. The military headquarters, established by Colonel John Graham on an abandoned farm, soon became a colonial settlement.

It was against 'Graham's Town' in 1819 that the resourceful and powerful prophet Makana advanced, commanding the 10 000-strong *impi* of Chief Ndlambe. The previous year he had defeated Ndlambe's nephew Gaika, a protégé of the British, at the Battle of Amalinde and was confident that he could rout the white invaders too. As they marched, the Xhosa warriors chanted that they had come 'to chase the white men from the earth and drive them to the sea'.

Unfortunately for the prophet, the *muti* applied to protect his warriors from the white men's bullets proved too weak against the grapeshot spewed out from the British cannons. Colonel Graham ordered mounted troops to plunder the villages of the defeated Xhosa and to burn their crops. There are varying opinions as to who were the 'good guys' and who were the 'bad guys' of this war, but it is known that ownership of cattle and rights to grazing lands were the crucial issues – and history shows clearly who became landlord.

Makana gave himself up to the British in a futile effort to stop their plundering of his people's homes. He was sentenced to life imprisonment on Robben Island, but on Christmas day of 1820 organized an escape and helped row an overloaded boat to Bloubergstrand. When surf capsized the boat, all but Makana made it to the shore.

After their defeat at the Battle of Amalinde, Gaika's forces escaped with their cattle northwards into the Amatolas. Gaika's Kop, where the Xhosa chief fortified his position, overlooks the mountain hamlet of Hogsback and the three 'hogsback' crests. To the west, across the Devil's Bellows Nek and Katberg Pass, the serrated peaks of the Winterberg slice the sky.

The pink and yellow helichrysum that grow abundantly in these mountains are known as 'ipepa' to the Xhosa, who believe that the flowers harbour shades – the spirits of their dead ancestors. The guiding powers of these spirits may be released either in the smoke when the dry flowers are burned, or in the steam when they are boiled in water. Today the Amatola range, liberally carpeted with these flowers, peacefully harbours the spirits of its more troubled past.

Although the Katberg escarpment is the highest part of the range, it is the Hogsback that makes the Amatolas so memorable. Apart from its historical interest, the secluded yellowwood forests below Gaika's Kop have a fairy-tale, even a magical, atmosphere. In winter snow lies heavy on the conifer boughs, ice crackles underfoot and streams gurgle over the damp, mossy stones. At the Thirty Nine Steps Falls a stream staggers down the rock stairs and after winding behind Hogsback village, descends in gossamer wisps down the Swallow Tail Falls to Keiskammahoek in the parched Ciskei below.

The ferns and long strands of lichen that cling to the columnar trunks of the ancient yellowwoods hang in shades of green. Brambles, wild berries and vines clamber on the forest floor and flowers burst outrageously beneath them. Coloured fungi and soft carpets of moss coat the rocks and tree trunks along the streams where tubes of sunlight cut through the growth like laser beams. In these spellbinding woods the young J.R.R. Tolkien went exploring in his early boyhood and it was here that his imagination was fired with the elfin kings, dwarf lords and dragons that manifested as the sweeping fantasies of a Middle Earth kingdom in his trilogy *The Lord of the Rings*.

In contrast, to the drier west the Karoo mountains march from east to west in platoons of flat-topped ranges interrupted by the occasional peak. Where the Stormberg halts, the Boesmanshoek Pass leads up to the Bamboesberg plateau, before dropping in steps and turns to the long plains below. The Cape bamboo, after which the range is named, fringes the banks of the Sandspruit and its little dams. On the slopes above Sandspruit farm can be seen the scars of coal mines dating back to the 1860s, and it is rumoured that prospecting for

92, 93, 94. *A fierce raptor of the mountains, the black or Verreaux's eagle is the most noble and still one of the most common eagles. These birds love to swoop in aerial acrobatics and mating pairs can sometimes be seen clasping talons in mid-flight and tumbling earthwards. Although the eagle is a ferocious predator, the ubiquitous crows and ravens occasionally taunt it in flight or even while it is eating.* 95. *From Katberg, the Amatola valleys and foothills lie spread out in front of the crests of Hogsback. The Xhosa chief Gaika sought refuge here in 1818 from the armies of Chief Ndlambe and the prophet Makana.*

commercially viable coal deposits is currently being undertaken.

The view from the plateau stretches far out across the Great Karoo, where koppies disappear into a midday haze. In the Karoo sunsets form spectacular backdrops of oranges and soft lilacs through a spectrum to deep purple. Pairs of black eagles dart past to roost, the white 'Y's on their backs and white windows on their wings flashing in the dying light.

The black or Verreaux's eagle is a graceful flier and majestic mountain raptor that swoops along the cliffs to surprise unsuspecting dassies, klipspringer or young baboons. Despite unwarranted persecution, black eagles are still fairly common and mountaineers thrill to see pairs somersaulting in aerial courtship and plummeting, with talons clasped, against the sky.

South-west of the Bamboesberg, the Bankberg exemplifies a transitional zone between the grasslands of the eastern mountains and succulent veld typical of the arid Karoo. Rocky areas are embroidered with grasses such as suurpol and turpentine grass, while thorny currant, renosterbos and the occasional wild olive grow scattered among the grasses. Near Cradock, at the eastern end of the rugged Bankberg's sweep, a species that came close to extinction has found a home. The Mountain Zebra National Park was proclaimed in 1937 with only six *Equus zebra*, but today their numbers have increased to well over two hundred. Soon after proclamation the park was restocked with animals known to have lived in the area before it was turned into farmland, and today visitors to this mountain sanctuary can expect to see eland, red hartebeest, black wildebeest, springbok, ostrich, steenbok, duiker and klipspringer. On the suurveld slopes and scrub-covered pediments of the south-eastern section mountain reedbuck and grey rhebuck may be flushed from cover, but the main predator of the park, the caracal or rooikat, is rarely seen as it hunts by night.

93

94

More than 200 species of bird have been recorded, including blue crane, korhaan, kori bustard (the heaviest flying bird in the world) and eagle owl, and the only pair of booted eagles known to breed in the protection of a national park are accommodated here. Less conspicuous but also abundant are cardinal woodpeckers, jewel-coated malachite kingfishers, hamerkop and numerous dry-country larks.

There are no perennial rivers in the park, although watercourses may gush in brown torrents after violent summer cloudbursts. But it does not rain often in the Great Karoo, and even when it seems that a wild summer storm is about to break upon the land, farmers watch in dismay as the heavy grey rain vaporizes in the hot draughts of ascending air before even one drop hits the ground. When summer rains do ease the land's unquenchable thirst, grass bursts green upon the hills that are flecked with soft flowers. During the winter months the Bankberg is cold and dry, and snow often powders the hilltops.

The stony hills harbour numerous species of lizard, skink and gecko, as well as other reptiles, among them the small East Cape adder, of which only twelve positive sightings are known. Many people have a revulsion for reptiles but of all creatures few are easier to befriend

an lizards; a little patience can turn
eir inquisitive nature into playfulness.
From the park the road to Graaff-Reinet
llows the southern edge of the Bankberg
d the Agter Sneeuberg. Beyond the
wn, the aptly named Valley of
esolation is flanked by the Camdeboo
ountains, the Sneeuberg and
ndjesberg, all remnants of older, higher
nd surfaces of the Karoo Supergroup
at have been worn back by prolonged
osion. In midsummer the wind across
e valley burns rather than cools, the
tense glare seems to squeeze the
eballs, and the afternoon disappears
to a shimmer of mirages. Travelling
orthward and following the rise of the

land to the Sneeuberg, the road to New
Bethesda unexpectedly enters a valley
where crisp mountain streams water
lucerne and cereal crops, fat sheep and
geese stare passively from the shade of
oak trees and willow branches sway
lethargically to and fro.

The village itself, sheltering in the
shadow of the giant, fanged Kompasberg,
offers cool respite on a summer's day, and
a luscious outspan. The Kompasberg
bears an uncanny resemblance to
Switzerland's great Matterhorn,
especially when it is coated with snow.
From its sandstone base sharply tilted
dolerite on the west face rises about
300 metres to the summit, 2 504 metres

96. *The dolerite fang of Kompasberg in the
Sneeuberg range dominates the landscape,
standing like a beacon above the hamlet of
New Bethesda. Its slopes have yielded some of
the finest fossils of the Early Beaufort period.*
97. *The Chinese lantern flowers in the Karoo
from July to September, illuminating small
patches of scrub. The attractive red-winged
capsules are the fruit of the plant.* **98.** *Blooms
of the Cape's Oxalis species are a sure
harbinger of winter.* **99.** *A fine mountain
zebra stallion sizes up the photographer while
his small herd disappears safely over a hill in
the Mountain Zebra National Park. Many
tribesmen will not wear the skin of this
animal, known as the rootless wanderer of
Africa, for fear of acquiring its ways.*
100. *Towers of volcanic rock on Valley
Mountain overlook the Valley of Desolation.*

100

above sea level. This is one of South Africa's least known mountains, yet is the highest point in the Cape Province beyond the Drakensberg basalts and it dominates the skyline like no other peak.

The southern and south-eastern cliffs yield good rock climbing, while the northern slope offers a scenic but tiring hike to the summit. Most of the rock climbing on the Kompasberg has been done by the Maclennan family – father Don and sons Ben, Joe and David – of Grahamstown. From the Settler City, Don Maclennan writes:

North of here are mountains
Which the sky has ground
Into a set of carving knives.
Walking on them you will find
The blunted edges pocked
With pink and yellow helichrysum –
Everlasting flowers,
Spirits of the dead.

The Kompasberg is surely the sharpest 'carving knife' to have evoked the imagery of the poem. On one of our numerous trips through the Karoo, Clive and I put a pleasant new F2 route up the south-east arête, solo climbing the easier, first three pitches.

Early travellers such as Le Vaillant and Governor Joachim van Plettenberg ventured this far east in the late 1700s and, although records are scant, it is believed that the first farms were established here not long afterwards. Since the settlers tamed their new land, tall, spreading oak trees around the farmsteads of Wellwood and Doornberg

101. (Previous page) *The arid Karoo waits for rain with long-suffering patience as storm clouds swirl over the mountains near Aberdeen.* **102.** *Fern-like Thinnfeldia grew abundantly in the Karoo during Triassic times, about 180 million years ago.* **103.** *To explain to a sceptical farmer how a certain fossil came to be on his land, pioneering geologist Andrew Geddes Bain said it was the remains of a creature on Noah's Ark that had disobeyed Noah, jumped overboard and sunk into the mud where it died. This skull of Cynognathus from the Early Beaufort period shows mammal-like features developing in the reptiles of the time. Canine, incisor and post-canine teeth have developed to increase the efficiency of mastication and thus help maintain a high metabolic rate. A temporal opening allows for bulging muscles while chewing.* **104.** *Thrinaxodon, from the Middle Beaufort period, is considered to be the earliest proper ancestor of mammals. Fossils of the same species were found in the Antarctic in 1976, giving substance to the theory of tectonic plate movement.* **105.** *The Cape rock agama is one of the most common reptiles in the Karoo and certainly one of the most good-natured.*

102

103

104

ound Kompasberg have seen the passing many generations.

As the southern slope sweeps sharply oward, the landscape changes abruptly. ove the fertile valleys that form a green irt around the base of the mountain, t-running spurs seem to have contour es drawn around them where quartzite s been exposed and these yellow ribs centuate the vertical dolerite columns. inty lydianite, metamorphosed between e sedimentary layers and the volcanic ck, was used by Stone Age man to shion implements.

The less steep northern slope asonally becomes a garden of wild eraniums, colourful watsonias, fat red ikerbos fingers and spreads of mauve uschia mucronata. Like Blouberg in the orth-western Transvaal, the Kompasberg an example of a mountain 'stepping one' for the migration of species, an land of fynbos and wild flowers in an cean of semi-desert.

105

A departing view of the mountain at unset, its tip concealed in low nimbus louds, with sunlight breaking through to olish the surrounding countryside, is nforgettable: burnished green flanks, heir ridges beaten with gold leaf where he light catches tall grass tips, purple ynbos shoulders, another green band and hen the slate-grey peak bites into the illowing cloud.

Less imposing but a more familiar andmark, Spandaukop stands guard over raaff-Reinet and behind it Valley Mountain commands a spectacular view ver the Valley of Desolation. The vista ramed by columns and tall pillars of olerite leads the eye across the plains of Camdeboo further into the Karoo's lryness, towards Beaufort West and aingsburg and the flat section of national oad that modern travellers speed across, ager to reach the embrace of the distant lex River mountains.

Travelling in the opposite direction, the

pioneers who drove their wagons over the plains of Camdeboo also saw it as a wasteland. Recent discoveries there, however, have led different eyes to see the Karoo as one of the truly great natural wonders of the world. For within its blankets of sandstone and shale are enshrined more than 50 million years of unbroken fossil records.

Time has been frozen in these rocks and today they form one of the world's finest palaeontological libraries. As it is exposed to the surface, each thin layer of rock describes a chapter in the story of evolution. For the marshy basins and windswept deserts of the Karoo's geological periods were the home of a succession of creatures that marked the development of life on earth, as it evolved from amphibians right through to the first mammals. Before delving into the story, however, we must briefly examine the geological make-up of the Karoo Supergroup and imagine what the land actually looked like during the various phases of deposition.

Rocks of the Karoo Supergroup cover the greater part of South Africa and nearly all the Cape Province north of its folded mountains. The four major deposition cycles of the Karoo period spanned about 100 million years: from the first ice-bound Dwyka phase during the Carboniferous period, about 250 million years ago, through the Ecca and Beaufort stages and finally to the volcanic cauldron of the Stormberg during Jurassic times, about 150 million years ago.

When the massive ice sheets of the Dwyka phase melted, the sludge left behind consolidated into rock known as Dwyka tillite that forms a ring around South Africa and parts of Namibia and Botswana, avoiding only the south-western Cape and far northern Transvaal. Dwyka rock has revealed many early marine fossils, including some from Namibia that appear to relate to ones

found in Uruguay. But the ice, stone and mud of the Dwyka glacial period was generally not favourable to life and the oldest animal fossil from the Karoo Supergroup is the 45-centimetre aquatic reptile *Mesosaurus*, found in the lowest shales of the Ecca Group, the period following the Dwyka phase, about 240 million years ago.

The landscape vacillated from marine to lacustrine, to swampy, to desert and back to lacustrine, leaving behind the Ecca shales and sandstones which bear few vertebrate or invertebrate remains. As well as *Mesosaurus* there is some evidence of marine life, but more typical of the Ecca period are the *Glossopteris* plants with their tongue-like leaves. Flowering plants had not yet appeared and plant reproduction was accomplished by primitive male and female fruiting bodies. Large fossil tree stumps and silicified tree stems have also been uncovered. It is the carbonized plant remains, however, that make the Ecca Group so important from an economic point of view, for they form the rich coal deposits in Natal and the Transvaal.

Only during the next period, when the shales and mudstones of the Beaufort Group were laid down, did terrestrial life in the Karoo begin to flourish. Swamps covered the ground and mists hung low over the primordial vegetation. The wet conditions were ideal for aquatic reptiles such as *Mesosaurus* to crawl onto land and there develop a new way of life. Thus began the great age of reptiles that culminated in the reign of the giant dinosaurs in the final chapter of the Karoo period.

The plant forms of Ecca times made way for early ferns. Gymnosperms, derived from fern-like pteridosperms, were represented by short, barrel-shaped cycads with crowns of leaves. Dragonflies with wingspans of up to a metre plied between the plant tops, and primeval cockroaches scuttled along the ground. Fish and amphibians swam in the swamps and open waters while *Euparkeria*, ancestors of crocodiles and birds, lay sluggishly in the rivers. Fossils show that by this time lizards were already a distinct group. It was during this period that the land movements to the south heralding tectonic plate movements that sheared Gondwanaland into the continents we know today, began forming the Cape Fold mountains, and drainage towards the interior increased.

During the period of the Molteno Formation, following the Beaufort Group, temperatures began to rise and rainfall decreased. With the climatic shifts the floral assemblage, like the animals, underwent many changes. A few species of coniferous tree were represented during Molteno times, as well as a similar order of Ginkgoales. These trees grew to a height of 25 metres and had wide-spreading branches and deciduous leaves. Today a single member of the order survives, in the ginkgo, or maidenhair tree. Fern-like plants such as *Dicroidium* became prominent during what is thought to have been deltaic conditions.

The next sequence, the Elliot Formation, probably consisted of floodplain deposits as the watery expanses continued to retreat, the mists lifted and the rains slowly withdrew. Plant fossils from this period are rare but the dinosaurs seemed to have thrived under the harsh conditions. Near Leribe in Lesotho can be seen the print that a monstrous three-toed reptile left behind as it bounded across the reddish silt.

Further and further the waters retreated. The rains became intermittent and for long periods ceased altogether. The plants withered and died as the sun beat down on the barren earth. Dinosaur skeletons must have lain in jumbled profusion, bleached and sandblasted upon the aeolian grit plains. The wind-borne sands sifted layer upon layer for the last, long, dry millions of years of this cycle, packing down into the sandstones of the Little Berg and eastern Orange Free State.

This layer, the Clarens Formation, forms a prominent ring of sandstones below the vertical Drakensberg basalts: its entire circumference is pocked with caves and overhangs that show evidence of Bushman habitation. Some are so deep that a number of farmers in the Barkly East district use them as shearing and milking sheds, and even barns. Unfortunately this expedient use of the shelters has destroyed many fine Bushman friezes.

The dramatic volcanic activity that produced the Drakensberg mountains marks the culmination of the Karoo Supergroup. While in the east lava gushed out on top of the sandstone to form basalt, in the west earlier volcanic intrusions solidified beneath the earth's surface as doleritic sills and dykes.

These horizontal and vertical bands of resistant dolerite form the familiar flat-topped tablelands and koppies of the Karoo where they have been exposed by aeons of erosion. Many still lie beneath the ground, waiting with the patience of eternity to play their part in decorating the country's ever-changing face.

At the time when the Drakensberg lavas were giving southern Africa the biggest fireworks display it had ever seen, albeit to an unappreciative audience, giant dinosaurs still dominated the land. A *Brachiosaurus* fossil found in East Africa shows a leg bone about two metres long, with a joint as big as a large watermelon; its owner must have weighed as much as 100 tonnes. But as vast streams of lava poured out of fissures in the ground, all the animal and plant life within reach was destroyed.

Fossils were first recorded in the mid-nineteenth century by the doyen of South African geology, Andrew Geddes Bain, who identified two general types: 'bidentals' and 'monsters'. It was then generally believed that fossils were organic embryos of creatures that, in their struggle for life, had failed to free themselves from the elements that gave them birth. Different evolutionary stages were described as varying degrees of embryonic development within the rock.

The story of modern palaeontology in South Africa begins in 1934 when, while on a picnic one afternoon, farmer Sidney Rubidge discovered that he had been sitting on a massive fossilized skull. At the time Dr Robert Broom was the country's expert in anthropology and palaeontology and the skull was despatched to him at the Transvaal Museum for identification. Spurred on by his find, Sidney Rubidge began to collect fossil remains and before long became a respected amateur palaeontologist. A cottage on the Rubidge farm Wellwood, in the shadow of the Kompasberg, now houses many showcases of expertly presented fossils, probably the finest private collection in the country. Many species and a genus, *Rubidgea*, are named after the man who helped open the world's eyes to this great palaeontological wonder.

Palaeontologists have few clues to work with when they try to piece together the enormous jigsaw puzzle of the earth's early inhabitants, but even the clues they do have reveal enough to see a progression of life right to the appearance of man. The Rubidge fossils in particular, ordered into family groups, show clear examples of mammalian features developing in the Karoo reptiles.

It is now generally accepted that animal life began in the oceans, where temperatures are fairly constant. The amphibians that first crawled out of the seas were cold-blooded animals, and like *Mesosaurus* and all later reptiles, could not maintain a constant body temperature. Once on land they were sluggish in cold environments and active only when warm. The first Karoo reptile appeared about 240 million years ago and formed an ecologically balanced fauna of both herbivores and carnivores, although the former were more numerous.

From Ecca times there were four distinct groups of mammal-like reptiles with varying degrees of mammalian development. Re-orientation of limbs improved locomotion; a sturdier skull housed a larger brain; and a stronger lower jaw with a temporal opening to allow for bulging muscles improved chewing techniques and increased feeding efficiency so that a high metabolic rate could be maintained.

The ability to breathe while chewing is
so essential to an increased metabolic
rate and this was achieved by the
development of a closed palate, which
shifted the internal opening of the
respiratory passage to the back of the
mouth. Another important development
was the division of the teeth into those for
nipping, biting and chewing, with the
formation of cusps on the teeth for
efficiently breaking down the food before
swallowing.

In late Karoo times the advanced
mammal-like reptiles were small
creatures that constituted an insignificant
group among the reptilian animals.

Dominated by the arrival of dinosaurs,
they did not survive beyond the Triassic
period, some 200 million years ago, but
their descendants radiated into a number
of mammalian families that came to the
fore with the dawning of the Tertiary

106. *Morning light strikes the dolerite crags of
the Valley of Desolation. Beyond the valley
stretch the Plains of Camdeboo, whose name
derives from the Hottentot word for 'green
hollow' and has been passed on to the
Camdeboo, or white, stinkwood trees that grow
in the mountains nearby.* **107.** *A rare Ruschia
mucronata flowers on the summit of
Kompasberg.*

107

106

period about 60 million years ago. In comparison, the earliest hominid fossils date from about three million years ago – a mere flicker in the light of this long evolutionary process.

This journey through the mountains of the Cape interior has taken us from the Stormberg through the layers of the Karoo Supergroup to the Dwyka formations, and it is here, at the western limits of the Dwyka deposits, between Beaufort West and Laingsburg, that our sojourn is completed. For about 200 kilometres the national road runs south of and parallel to the distant ranges of the Nuweveldberg and Komsberg, whose dolerite-topped hills bounded by semi-desert are typical of the Karoo. South of the road, however, the mountains of the Witteberg are not flat-topped, but rolling, cut by numerous kloofs, and in places they appear to have buckled.

The most northerly range of the Cape's folded mountains, the Witteberg is made up of rocks of the Cape Supergroup, the cycle of deposition which preceded that of the Karoo. But whereas the Karoo mountains have been formed through peneplanation, or levelling, it was the tremendous force generated by movements of the earth's crust that folded and twisted the Cape sediments into the belt of mountains that runs parallel to the coastline.

108. *Rounded sandstone boulders stand imposingly in the Mountain Zebra National Park. The park was proclaimed in 1937 as a refuge for one of the few small herds of mountain zebra that still roamed the Bankberg.* **109.** *Aloe striata brings a splash of winter colour to the rocky hillsides.* **110.** *Delicate white flowers crowd at the tips of Euphorbia esculenta's green fingers to attract the pollinators' attentions.* **111.** *The overhangs along the Bell River sheltered Bushmen for thousands of years. Had any survived into the twentieth century, the sandstone valleys around Rhodes village and Barkly East might have been an ideal sanctuary for them.* **112.** *(Overleaf) Vultures are no longer seen around Aasvoëlkrans near New Bethesda – not even tell-tale white streaks of guano on the rock are left as evidence that they once nested here in abundance. The feathery row of introduced poplar trees in the foreground is an example of man's encroachment on natural habitats.*

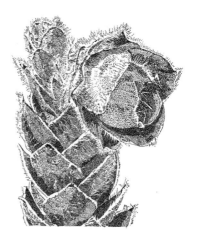

THE CAPE FOLD RANGES
MEETING-PLACE OF MEN AND MOUNTAINS

Sinister rise the mountains, jagged and
  bleak and bare,
Cloven and rent and fissured by fire and
  torrent there;
But the moon is a tender lady that loves
  not sights like these,
And in her spell transfigured, all things
  must soothe and please.

John Runcie
Crossing the Hex Mountains

To the early white settlers of the Cape, the mountains that cut them off from the hinterland did indeed seem 'sinister' and 'jagged'. But these barriers also kept the forboding savagery of 'darkest Africa' out of mind and, as long as the green valleys of the south-western Cape provided all they required, only the most curious and courageous of the settlers dared to cross them.

Today even South Africans who have never visited the Cape mountains feel intimate with them through the paintings they have inspired in artists over the centuries: the white Cape Dutch farmhouses surrounded by red, green and golden vineyards, with their backdrop of high mountain crags and kloofs. The western Cape and Boland is a classic meeting-place of man and mountain where awesome ranges, fertile valleys and a relatively benevolent history have given rise to an organic culture that defies the harsh continent.

From this corner the Cape Fold ranges run northward to the Cedarberg and eastward to the Grootwinterhoekberg near Port Elizabeth. Like many of the world's greatest ranges, they occur as elongated belts parallel to the continental margin and are the result of intense upward displacement of the earth's crust. The actual cause of the folding on our

113. *A lump of oxidized iron ore offers itself like fossilized fruit in a lichen-covered sandstone bowl.* **114.** *A climber forges his way up a series of overhangs in the Cedarberg. The Cape mountains have the most multipitch, high-grade climbs in southern Africa.*

subcontinent is not as clear as that of the Alps or the Himalayas, which are the crumpled panelwork resulting from the collision of tectonic plates.

However, geologists believe that the formation of the folded mountains began with crust movements that caused deep downwarping along the southern edge of the continent. The bowl-shaped depression thus formed became the repository for thick layers of sediment of

the Cape Supergroup, and the great weight of these, together with the downwarping, disturbed the lower layers of rock and squeezed together the lips of the depression. Consequently the strata surrounding the region were folded, and is the eroded remnants of these folds that can be seen in the southern and south-western Cape. The twisted and tightly buckled rock in the Outeniqua, Swartberg and Hex River mountains bears testimony to the tremendous force that created these ranges.

Sandstones of the Table Mountain Group not only form the oldest layer of the Cape Supergroup, but are also its most visible component. Weathered to greyish white or reddish-brown, they are easily identified as the peaks of the Outeniqua, Groot Swartberg, Riviersonderend, Hex River and Du Toit's Kloof mountains, as well as the upper section of Table Mountain itself.

Shales and sandstones belonging to the Bokkeveld Group form the next oldest layer and underlie many of the Cape's beautiful, fertile valleys. The marine brachiopod fossils found in these strata suggest an age of about 300 million years, which coincides with the great era of fossils in Europe. In our subcontinent, however, relatively few from this period have come to light, although the Bokkeveld rocks have revealed a number of trilobites and two types of gasteropod.

The Witteberg sediments, the youngest of the Cape Supergroup, form the prominent crests of the Swartruggens near Ceres. Generally harder than those of the Table Mountain Group, these rocks are more red or yellowish and weather to white. They have yielded mainly plant fossils, but also the worm *Taonurus* and a number of fossil fish.

Today's folded landscape changes

considerably both in its approximately 800-kilometre span from east to west, and from north to south. Picking up our mountain spiral from the Amatolas in the east, the folded ranges offer a contrast between the green forests of the most southerly ranges and the rocky scrub hills, coarse suurveld and Karoo plains of their parallels to the north. Within a few kilometres the climate can change from the zones of high winter rainfall to the Little Karoo where cold, dry winters are harsher even than the equally dry, sweltering summers, in a land fit only for goats and reptiles.

In the east the Grootwinterhoekberg forms the first significant range of the Cape Fold series. It is known locally as the Cockscombs, for the main outcrop near Steytlerville has five closely connected crests that resemble a rooster's comb. The crests are the main climbing attraction of this region, where 400-metre faces yield good climbs. However, eastern Cape climbers are beginning to look more closely at the tall dolerite crags further inland for challenging rock climbing.

In the Kouga mountains south of Willowmore, Prinsloo's Kloof sets the scene for one of the most gruesome of the many ghost stories that shroud the Cape mountains in mystery. Jan Prinsloo was a cruel man who so mistreated his Hottentot servants that one day they turned on him and in a frenzy butchered him. Since then no owner has kept the rich property for more than a few years and it is shunned as a place of evil, where replays of the macabre attack take place on each anniversary.

Some of the most interesting areas of the Cape Fold mountains have been made accessible to hikers by a number of trails, of which the longer ones form part of the National Hiking Way System. The Tsitsikamma Hiking Trail, between Nature's Valley and Storms River, runs parallel to the popular Otter Trail but follows the base of the Tsitsikamma mountains. It meanders through high, humid indigenous forest on shale, of which one of the most splendid displays is at Nature's Valley, as well as flowering Cape fynbos on quartzite soils.

In the Lottering State Forest the trail passes Formosa Peak (1 675 metres) which stands at the head of the Bloukrans, Krakeel and Louterwater valleys. Sparse fynbos grows in the rain shadow of the northern slopes, while the lush southern slopes are covered in dense forest and thick fynbos.

During winter, mists from the sea roll across the range and snow often falls here. Strangely though, this is also the season when berg winds sweep down from the interior to the sea, causing high temperatures in midwinter. Storms are frequent in summer and hikers are sometimes caught unawares when the fast-flowing rivers come down in flash floods.

Inland of the lush Garden Route of forests and lagoons, beyond the wrinkled mountains and valleys, the Groot Swartberg borders the Little Karoo. As one approaches the mountains, it is difficult to imagine where the road could possibly traverse these wild heights, or to believe how Thomas Bain – son of Andrew Geddes Bain who engineered Bain's Kloof Pass – could have built a road that runs 1 200 metres above the Prince Albert valley. But the brooding cliffs do give way, and the Swartberg Pass snakes over the mountains and descends to the Little Karoo. The old toll house, built originally to house Bain's convict labourers, now stands in ruins and is believed to be haunted.

At the western end of the Groot Swartberg, where it is cleft by the Gamka River, is the lonely Gamka Kloof, discovered about 200 years ago by nomadic trekkers who called the valley 'De Hel'. For more than a hundred years

they lived there, isolated from the rest of the world, until 1921 when a small school was built in the valley. In the early 1960s a road was cut through the fields and wooded slopes and soon the outside world began to encroach on the valley's serenity, but De Hel is still a strange and solitary place.

Further west, in the Klein Swartberg, 2 330 metre-high Toverkop dominates the area around Ladismith and is easily recognized by its twin towers. Although mountaineering in this country was born, naturally enough, on Table Mountain, it was on Toverkop one Sunday in 1885 that the story of rock climbing begins. Nineteen-year-old Gustav Nefdt led five

115. *Two-million-year-old scallop shells, these marine brachiopod fossils were found in the Bokkeveld deposits, one of the three main geological groups that make up the Cape Fold mountains.*   **116.** *The rare Fockea crispa is known in the Little Karoo as the 'ghwarriekoe' as it grows under and is camouflaged by the ghwarrie bush. During the nineteenth century a specimen in the Schönbrunn Gardens in Vienna was thought to be the last survivor of a species that, in fact, still grows in the Swartberg near Prince Albert.*
117. *Cockscomb's large crest near Steytlerville attracts eastern Cape mountaineers to the Grootwinterhoekberg.*

115   116

117

ther young Ladismith men in secrecy up ne mountain. None had any climbing xperience and, without climbing quipment, only Nefdt managed to scale ne western tower.

The townsfolk would not believe their tory, declaring after a reconnaissance of overkop that it could only be scaled by zards, and they were not far wrong. To indicate his honour Nefdt once again ackled Toverkop's face and by using aybacks, hand jamming and other echniques of the modern climber, he gain stood on top of the mountain. He nen had to lower a rope to bring up the est of the party.

To put this amazing achievement in erspective, Nefdt's route up the verhanging crack was attempted many imes by generations of the country's best limbers. It was not until 1947 that he was inally followed by Harry Currey, Brian Russell, Ted Keen and Denys Williamson. After an unsuccessful attempt in 1931, amous Cape climber Bert Berrisford said: Toverkop is an extraordinary mountain, nd in many ways an appalling and errifying one. . . Where one expects o find a saving grip none exists - the mountain seems to hold out no riendly hand.'

The Outeniqua range, between the Swartberg and the sea, boasts a 148-kilometre trail through indigenous forest nd montane fynbos that takes eight days o complete. Although it is one of the most rduous of the national trails, the year-round rainfall in the area makes it very ewarding for amateur botanists. Lucky hikers may see elephant spoor at Rondebossie, but are highly unlikely to see any of the few Knysna elephants that still roam the forest. At Millwood the hundred-year-old ruins of a short-lived gold rush are enshrouded in vegetation.

The longest range of the Cape Fold mountains curves gently for some 200 kilometres between the Gouritz and the Breede rivers. The Langeberg has a

softer aspect than the rugged Boland mountains, with grass and farmlands rolling well up the lower slopes. In the Boosmansbos Wilderness Area, roughly in the centre of the range, hikers can explore the slopes of the Grootberg or climb to its summit. The forest reserve in the Wilderness Area is one of the country's larger remaining patches of Cape indigenous forest and contains, among others, mountain cedar, stinkwood, yellowwood, beech and Cape holly. The bright red fruit of the moisture-loving holly is favoured by many birds, particularly the Cape bulbul and red-winged starling. Sunbirds are also frequently seen along the trail.

One of the most beautiful of the National Hiking Way System trails is the Swellendam Trail, on which hikers follow a circular route around the main peaks in the Marloth Reserve. The mountains in this part of the Langeberg form prominent peaks and rocky ridges with forested ravines, and in early summer are carpeted with pink ericas. Misty Point, at 1 710 metres the highest peak of the range, is usually climbed only by mountaineers, but Twelve O'Clock Peak is accessible to hikers, who can follow a zigzag path to the summit. The trail offers inexhaustible possibilities for exploration in the ravines and forests, and some of the most picturesque scenery in the Cape.

Between Stellenbosch, Franschhoek and Grabouw, in the Hottentots Holland mountains, the Boland Trail also offers a number of long and short hikes through beautiful fynbos country. However, capricious weather is a hazard for unprepared hikers. Mists come down without warning and cold, wet and windy conditions have claimed several lives through hypothermia. In addition, heavy winter rainfall can make some of the paths impassable.

There are three main options when tackling the Boland Trail. My favourite route passes Somerset Sneeukop (1 590 metres), crosses the Lourens River and continues past the Triplets into Swartboskloof. A worthwhile detour to the east of Haëlkop leads up the Eerste River to a spectacular waterfall. The next section, through the Jonkershoek State Forest, is the most beautiful I have seen; from the path the view of The Twins (1 504 metres), Ridge Peaks (1 515 metres, 1 516 metres and 1 517 metres) and Banghoek Peak (1 526 metres) is magnificent. While the hiker may marvel at the attractive fynbos, refreshing mountain pools and vibrant birdlife, the summits of the peaks can be gained only by well-equipped mountaineers.

120

A detour down Boegokloof emerges at its junction with the young Rivier-sonderend, the 'river without end'. The trail veers to the north-east, towards Aloe Ridge and views of the luscious Franschhoek farmlands, but for those prepared for adventure, the Rivier-sonderend canyon is waiting.

Exploring canyons, known as ' loofing', is a popular mountaineering pastime and a kloofing trip in the Cape Fold ranges is hard, but rewarding. It includes boulder-hopping, very likely some awkward scrambling on slippery rock and definitely swimming – intentional or otherwise.

A waterproofed pack and inflatable raft make the journey more bearable, and a small stove is essential for preparing a hot drink after a freezing swim. One of the most exciting kloofing experiences is up the Elands River in Du Toit's Kloof, where riverine vegetation hangs down the red sandstone walls. A swim in the big, dark pool to the waterfall at the end of the kloof is daunting, but unavoidable. Unfortunately this kloof is often badly littered, but Mountain Club members have access to some of the more remote places in Du Toit's Kloof, as well as in Bain's Kloof and the Hex River mountains.

Legend has it that the Hex River valley and mountains get their name from the bewitched spirit of a young beauty named Elize Meiring. She sent a suitor up the

118. The rocky slopes of Baviaanskloof in the Little Karoo offer ideal conditions for the tall euphorbias that flourish there.

119. A cutaway cross-section of a mountain reveals the power of the geological forces that formed it, solidified in its contorted strata. It is estimated that these folds were pushed up to a height of 6 000 metres before erosion ground them down to the bases we see today.

120. The Aloe ferox or bitter aloe not only features in prehistoric Bushman paintings but is still an important species. The leaves provide food for stock animals in times of drought, they make excellent jam, and the laxative drug 'Cape Aloes' is obtained from the plant.

nearby Matroosberg to fetch a red disa from the highest point of the mountain and so demonstrate his love. The mountain claimed his life, as it has the life of many a heedless youth before and since. The young woman died from grief and remorse and her spirit now roams the misty, gusty range.

The Hex River valley, where De Doorns was once a far outpost of the early colony beyond the mighty portals of Du Toit's Kloof, has the reputation of being the most beautiful in the country, yet the mountains, too, are impressive in their power. Looking southward from the Hex River Pass across to the Kwadousberg, the eye is drawn up by the land's sweep as it curves into the huge cavity in the mountain. Bands of yellow and orange sandstone flow in waves into the great amphitheatre, one of the best examples of folding to be seen anywhere.

Amongst all this mountain chaos is a calm centre on which all lines seem to converge. Surrounded by the Hex River, Klein Drakenstein, Du Toit's and Slanghoek ranges, the Boland town of Worcester stands in a wide valley, and when the fierce Cape winds tear through the mountain passes it seems to be the sheltered eye of the storm.

From this centre the road travels deeper and deeper into Du Toit's Kloof and closer to what seems to be a solid wall of high mountains, then suddenly slips between the peaks. It twists sharply up the valley to the top of Hawequas, and from there looks down to the round granite 'jewels' of Paarl in the valley far below and, beyond, to Table Mountain, 50 kilometres to the south-west.

In the middle of Du Toit's Kloof stands Milner Peak, another great natural amphitheatre whose face was long considered impossible to climb. In December 1946 a party of four led by 'Shippy' Shipley and Jannie Graaff overcame the extreme exposure of its 600 metre frontal overhangs. 'Fifty feet of climbing on minute holds over an awesome drop enabled us to overcome

this obstacle and, sitting on a small ledge at the top (of the crux pitch) with our legs dangling over the void, we scratched our heads over the problem of "where next?"

'The next pitch was the same – incredibly exposed and frightening but, once we had plucked up the courage to get started, quite a reasonable proposition,' recounted Shipley.

In 1967 determined pioneers of the western Cape mountains Hans Graafland and Mike Scott made a Christmas weekend first ascent of the right-hand buttress. 'The magic spell of the Hex River mountains is surely cast by monolithic Milner Peak. In an incredible surge of rock, upthrust like Atlas' shoulders to prop up the sky, the peak stands guard with the bludgeon of Milner Needle at its side.

'"Fantastically beautiful" is the only way to describe it when the mists swirl around the huge orange walls. Twin flying buttresses at the base suspend aloft these sweeps of cinemascope screens, petrified into a heliograph that flashes to climbers a lure and a challenge.'

Two years later, with the formidable Tony and Robin Barley, Mike Scott climbed 'Apollo' on the central buttress of Castle Rocks, a face that had eluded strong climbers over the period from the Apollo 8 moonshot until four Apollo space trips later. Scott and his companions accorded it the title of 'Last Great Problem' ('LGP'), the ultimate rock climbing route of the day, and their conquest stirred other Cape cragsmen into a frenzied search for new

121. *From the Langeberg, the gnarled fingers of the Attaquas mountains point the way to the Outeniquas in the far distance.* **122.** *Bracken and fynbos carpet the foothills of the Outeniqua mountains.* **123.** (Overleaf) *Toverkop near Ladismith in the Klein Swartberg was the first serious rock climbing obstacle in the country to be tackled. In 1885 young Gustav Nefdt surprised himself, his friends and the townsfolk by solo climbing the tower up an overhanging crack system that for the next 70 years repelled the country's finest climbers.*

122

124. *Part of a continuous bed of folds, these parallel bands of rock have been vertically uptilted by titanic forces within the earth's crust and then ground down into serrated knife edges.* 125. *Shy spring-flowering members of the iris family allow only the sun's warm fingers to coax open their petals. This species occurs at the top of the Swartberg.* 126. *Proteaceae are the most visually striking family of the fynbos plant kingdom. While some members are widespread, many are confined to specific ecological niches in small areas. Leucospermum oleifolium is found in several ranges of the south-western Cape.* 127. *A gulley in the Seven Weeks Pass in the Groot Swartberg shows folded bands of sandstone that characterize the entire folded mountain system.*

126

7

GPs'. Over the next ten years climbing barriers were pushed back fast and furiously, new generations of climbers were spawned, and routes with unimaginable grades were attempted.

Antonio de Saldanha was the first man known to have reached the top of Table Mountain, and since he did so in 1503 the country's most famous landmark has attracted climbers of all kinds. By the 1790s rambling up Table Mountain was a popular pastime but the first man to become closely acquainted with it was the unlikely Joshua Penny, RN.

An American, Penny was pressganged into the Royal Navy and deserted HMS *Spectre* in Table Bay in 1799. To evade capture he took refuge among the Peninsula's mountains and spent 14 months living off the land. He lived near a spring in Fountain Ravine, killed buck with stones and wore their skins. He even made crude mead in a hollowed-out tree trunk. 'I never enjoyed life better than when I lived among the ferocious animals of Table Mountain,' he said. Buffalo, lion, elephant and leopard, in fact many of South Africa's wild species, were first seen where the Cape Town metropolis now hugs the mountain.

For the past two centuries Table Mountain has been the centre of this country's climbing activity, but with the founding of the Mountain Club of South Africa in 1890 exploration of other areas of the western Cape mountainland began with great enthusiasm. Sepia photographs from the Edwardian era show large mountaineering parties assembled in front of railway coaches for country outings to the Hex River, Winterhoek and Drakenstein.

The first climber of note in South Africa was George Travers-Jackson who, from its inception, was an active member of the

128

129

128. *'The people who sit on their heels', as the Bushmen call baboons, are the most versatile creatures of the wild. They may be found high on mountain crags, deep in forests or on the hot plains of the Bushveld.* **129.** *The rooipypie, Anapalina caffra, flowers from September to December in the mountains of the southern Cape.* **130.** *Mountain reedbuck are more numerous than their retiring manner suggests. Groups of two or three may be seen on grassy mountainsides but they are quick to flee if approached.*

Mountain Club for more than 50 years. He quickly established a high standard of mountaineering and an adventurous attitude to route finding. Then came Ken Cameron, the Berrisfords and the brilliant George Londt.

In about 1920 Londt became the leader of an energetic climbing group that opened many new climbs on Table Mountain. He was a short but powerful man and because he did not have the reach of other climbers, made up for this limitation by becoming a great innovator of climbing technique. In 1920 he led the first frontal attack on the Klein Winterhoek, a magnificent peak above Tulbagh that has 400 metres of climbable rock and is now a classic country route. Then, on 13 November 1927, the herculean, meticulous Londt, who had never put a foot or hand wrong, fell to his death on Table Mountain. His only mistake was his last, and the argument as to how he could have fallen on a relatively simple climb continues.

One good leader tends to motivate others in a master-apprentice relationship. Travers-Jackson spurred on others and Londt masterfully coached his climbing colleagues in the finer art of rock work. After Londt came Frank Berrisford and other competent leaders to carry on where he had left off.

During World War II rubber sandshoes

became part of the equipment of climbers who intended climbing on 'thin' rock – a move that was bitterly resisted as unsporting by the hobnailed old school. As climbing progresses with new techniques and equipment, there is a constant conflict over ethics between some of the older climbers and the new adventurers.

Climbing in the Cape is glorious during the summer, but caution has to be exercised against dehydration and heat exhaustion. In winter freezing winds and snow may suddenly descend upon a mountain. Such was the case in the late winter of 1945 when a group of six set off for a skiing weekend at Waaihoek. The rescue party that was eventually alerted found three of the group frozen to death and the others alive but in a desperate condition.

The end of the war brought new technology to all spheres of life, and to mountaineering a piton that was both cheap and safe. It caused heated debate between the old and new schools, but had a profound effect internationally. By improving the safety of free climbing and allowing sophisticated aided ascents, it pushed back the barriers of the possible.

The 1950s saw a flourish in climbing activity and talent in the Cape, as in Natal and the Transvaal. It included the partnerships of B. Fletcher and D. Carter,

and M. Mamacos and C. Butler. Later the were joined by the unorthodox 'hardman Rusty Baillie, the first South African to scale the meanest of all peaks, the north face of the Eiger.

The number of G routes on Table Mountain increased rapidly and new classic routes such as the north-west face of Du Toit's Peak were achieved. In 1960 'not being a purist' Baillie pioneered aided climbing in South Africa and soon mountains that had been considered unclimbable were falling to the human flies. The modern tendency is to free climb previously aided routes, but mechanical assistance over short section: is no longer frowned upon as it once was.

For years Stellenbosch had produced her own mountaineers and in the late 1950s some of her progeny burst onto the national climbing scene. Most notable were André Schoon, Henri Snijders and Hans Graafland. During the 1960s Graafland teamed up with Mike Scott and for the next ten years they made a formidable team. Country routes up to 25 pitches long, many of sustained G standard, were credited to them.

In the early 1970s there was another rush of leaders in the Cape, most notably Dave Cheesmond, Richard Smithers, Ed February and D. MacLachlan. By the latter half of the decade the most surprising thing about the top rock climbers was their youth; teenagers were among the finest climbers in the country. Within a few years of their first ascents, mountaineers such as Dave Davies, Chris Lomax and Greg Lacey were on a par with the Barleys, Dicks and Cheesmonds – the climbing élite.

This upsurge in young talent was robbed of one of its brightest stars when Roger Fuggle died in a climbing accident in 1974. At the time the Natalian was enjoying a climbing season at Chamonix in France with his long-time partner Tony Dick. At the age of 28 Fuggle was already an esteemed mountaineer who had opened a number of classic routes, had been on major expeditions and made the second pitonless ascent of El Capitan, the greatest of all rock routes at Yosemite in the United States.

In 1977 Cheesmond, MacLachlan, Butch de Bruin and Tony Dick opened 'Dinosaur Revival' at Duiwel's Kloof in the Groot Drakenstein. This 15-pitch climb, including seven F3 pitches, six G1 and one G2 pitch with four aided moves, was achieved in 13 hours with a bivouac. Three months later Lomax, Lacey and Brian Gross made the ascent in nine hours, freeing all the aided moves except one on pitch 12. They graded three more pitches at G2 and pitch 12 a possible G3 – phenomenal for a long ascent.

That year heralded something of a climbing renaissance in the Cape and apart from 'Dinosaur Revival', Cheesmond and Dick opened up two more 'LGP' routes, fittingly named 'Renaissance' and 'Time Warp'. Of 'Renaissance' they said: 'The popular comment about "the most serious undertaking in the country to date" is probably applicable. The only difference is that we thoroughly recommend the route for enjoyment as well as seriousness, because of excellent rock.'

A great watershed in rock climbing was reached in 1979 and 1980. By the early 1900s Travers-Jackson had attempted climbs of G grade but it was only 50 years later that sustained G grade climbing became possible. By the 1970s G had been pushed up to G3 but only by three or four superb climbers. In 1979 Lacey and Dave Kelfkens opened 'Tour de Force' on Table Mountain, the first H route in the country. On the same mountain a year later Kelfkens and H. Davies climbed an H2 route they called 'Cool Cat'. Others followed.

Dave Cheesmond is currently South Africa's most esteemed climber on rock and ice, and has an international reputation as an expedition mountaineer of the highest order. Now based in North America, in 1983 he was a member of the successful United States Everest

expedition and has been asked to lead a Canadian team up the same peak in 1985.

This latest generation of climbers has begun a trend of solo climbing that once again has the old guard and young turks in furious argument. Before the days of 'nuts and slings and ropes and things' the popular form of rock climbing was for one party to follow another, without any aids, up easy climbs graded C and D. All subsequent moves away from this were considered unethical. Today, a number of local climbers have reverted to climbing individually without any aids – the difference being that they are now solo

climbing well into the G grades. They have been branded by some as egotistical and reckless, and that may be so. Most experienced cragsmen do a certain amount of solo climbing, usually on a route that has comparatively easy pitches which can be speeded up by abandoning safety precautions. Solo climbing an E route would not attract any attention; soloing a G route has mountaineers drawing metaphysical battle lines.

What drives these solo climbers to tempt fate so? They are superbly trained athletes and, like all other athletes, are concerned with improving their craft. In

31. *The Langeberg's 200-kilometre spine divides the fertile coastal plain to the south from the Little Karoo in the northern rain shadow.*

this context it may be argued that they are driven by competitive spirit. If this is so, they are transgressing the mountaineering spirit of comradeship that transcends competition.

On the other hand, in 1971 German mountaineer Reinhold Messner, without doubt the world's greatest alpinist, wrote of mechanical climbing: 'Today's climber carries his courage in his rucksack, in the form of bolts and equipment. . . Faith in equipment has replaced faith in oneself.' Perhaps today's rock gymnasts are trying to reassert their faith in themselves.

Climbers and hikers in the western and southern Cape mountains perhaps see the best that the smallest floral kingdom in the world has to offer. So richly diverse and unique is the flora of this region that it has been accorded the status of a full plant kingdom, one of only six worldwide and, with an area of only 18 000 square kilometres, by far the smallest. The Afrikaans word 'fynbos' was first used as a botanical term in 1916 and is now universally accepted to describe the characteristic macchia-type Cape vegetation. Divided into three major elements which are represented by the broad-leaved, woody Proteaceae, the minute-leaved ericas and the reedy Cape grasses or 'restios', fynbos has its epicentre in the western Cape where winter rainfall and long, hot summers are typical of a mediterranean climate, and where it thrives on nutrient-poor soils.

One of the enigmas of macchia vegetation types in mediterranean climates throughout the world is their similarity despite having unrelated origins. Thus, in a fine example of what scientists call convergent evolution, completely different species from the tr[u] Mediterranean, from California, Wester[n] Australia and the Cape all exhibit the typical leggy growth form and sclerophyllous leaves.

Another enigma to botanists is fynbos speciation: why are there so many different plants in the Cape fynbos biom[e] A selected square metre may contain dozens of different species, and on Table Mountain alone some 1 400 species occu[r] Moreover, many species, even whole

nilies such as the Bruniaceae, ubbiaceae and Penaeaceae, have olved only in the south-western Cape. deed, the large number of species demic to the fynbos was one of e reasons for it being declared a ral kingdom.

It also appears that the most beautiful d delicate fynbos flowers flourish der the harshest conditions, in the ghest and most exposed locations. In e mountains rainfall varies considerably d can reach 2 000 millimetres a year, curring mainly during winter when ow is also common in the west. The mmers are hot and long, but sometimes leviated by moisture-bearing clouds om the south-east. About 7 000 species

thrive in the mountains and half of these, including some 50 of the 600 erica species and 320 of the 360 proteas, are endemic, sometimes to very small areas.

The tea-coloured mountain streams, whose distinctive hue results from interaction between vegetation and soil types, are fringed with riparian thicket and a mixture of fynbos and scrub forest. If they receive prolonged protection from fire, protea bushes and other tall plants may form impenetrable thickets up to six metres high on the mountain slopes.

Fire, however, plays an ambivalent rôle in the survival of the fynbos. The vegetation is highly combustible and during the dry summers, when the South-easter blusters and the flames can be

almost impossible to control, large areas of natural vegetation are frequently devastated. This is beneficial in that the area is cleared of thick tangles of old vegetation and new growth is encouraged. Many fynbos species regenerate from a fire-resistant rootstock or from seeds or

132. *The crags of the Franschhoek mountains, one of five ranges that form a mountain vortex between Stellenbosch and Franschhoek, are well known to mountaineers. Extreme rock and weather conditions mean that only well-prepared climbers reach the top of these challenging peaks.* **133.** *A number of difficult climbs in Duiwel's Kloof in the Grootdrakenstein mountains were opened during the late 1970s, heralding a rock climbing renaissance in the Cape.*

133

bulbs that remain dormant underground until fire gives them the stimulus to germinate. In the absence of fires, plant communities may become choked with dead material and die off. The salutary effect of fire was dramatically demonstrated when the beautiful marsh rose, a member of the protea family, almost became extinct in 1968. The area in which a few single plants were struggling to survive was burnt, and 12 years later a population of more than a thousand was thriving.

Too much fire is as bad as too little and where fires occur too frequently, not giving the fynbos time to regenerate, much valuable vegetation is destroyed. Moreover, too much fire encourages the proliferation of fast-growing alien vegetation that is taking over the fynbos habitat.

Originally introduced to stabilize drift sand areas or for forestry purposes, alien wattles, pines and hakea have completely replaced the natural flora in some places, and pose a very serious threat to the survival of the fynbos. The prime offenders are hakea, the rooikrans wattle, the long-leaved wattle, the Port Jackson willow and the blackwood, all of which come from Australia, as well as the cluster pine from the Mediterranean. Strenuous efforts are being made to control the spread of these invasive plants, one of the more successful being volunteer 'hack groups' who spend their weekends cutting down the alien vegetation.

Cape Town University scientist G.N. Louw wrote in a study of the fynbos region conducted during the late 1970s: 'The Fynbos Biome has become so threatened in the past two decades that immediate action is required. For those who still need convincing, I only need draw attention to Table Mountain – one of our national monuments. It is eroded, littered with unsightly debris, disfigured by grotesque buildings, covered with alien weeds and has become an eyesore to the layman and a tragedy to the ecologist.'

But why conserve? The best answer is the brief title of the book *Extinction is Forever*. Already about 60 per cent of the fynbos area has been lost to farming, forestry and urban sprawl. Apart from the necessity of protecting all species, both rare and common, to maintain the stability of the ecosystem, there is a more general reason for conservation. It is one that most mountaineers would identify with, for it is akin to the aesthetic appeal that draws them to the mountains – the land ethic, to love and respect the wild places. The MCSA's seventh stated objective is 'to protect and preserve the

134

natural beauty of the mountains and the natural water supplies'. Mountaineers may be the most irreverent collection of individuals pursuing a similar recreation, but few would find fault with that objective.

The northernmost extension of the Cape Fold mountains is the Cedarberg, which shelters some of the rarest endemic fynbos flowers. On the highest peaks of this range the snow protea, whose scientific name *Protea cryophila* means cold-loving, is sometimes found. Its striking white centre is encased in pink petals whose tips seem to be flecked with snow.

The most attractive flowering plants, such as *Protea magnifica*, the bearded protea, occur in the high-altitude zone. The showy *Leucospermum reflexum*'s fire-burst flower is well known in town gardens but grows naturally only in the northern Cedarberg among the peaks of Krakadouw. Erica flowers of yellow and pink sweetly scent the mountain air, while in the valleys wild almond, water witels, silky cone bush and the lovely yellow marguerite may conceal grey rhebuck and Cape grysbok. Klipspringer and steenbok frequent the waboom veld of the rocky scree slopes. Although the waboom's excellent wood was much in demand, particularly by early settlers who

used it for wagon building, these gnarled tree-proteas are still a common sight in the Cedarberg. On the lower slopes grow rooibos and the round-leaved buchu used in flavouring a reputedly medicinal brandy. Both species are now grown commercially on a large scale.

The first white men to set eyes on the Cedarberg were probably members of a Dutch East India Company expedition in 1661, which Governor Simon van der Stel had sent northward in search of the legendary Monomotapa, the golden kingdom of central Africa. They did not get far, and headed back to Cape Town soon after one of the party had been killed by a lion at Piketberg.

By the early eighteenth century a way around the mountains had been established and farmers settled along the Olifants River valley's fertile course. Only the most intrepid travellers actually penetrated the range, men such as John Barrow and Martin Lichtenstein, and they wrote of the excellent properties of the wood of the cedars which grew in thick forests there and gave their name to the range. Over the next 80 years woodcutters settled in the mountains and by 1840 Sir James Alexander, when reporting on his travels, complained bitterly at the rate at which these magnificent forests were

being exploited and uncontrollably burned. In 1876 the first forester was appointed to the Cedarberg and conservation was gradually introduced.

There are now few cedars left in the Cedarberg and even under the Forestry Department's careful supervision it will take a long time before they are re-established in reasonable numbers, as the trees take many years to mature. Periodic droughts and fires continue to take their toll of the gnarled, weather-beaten trees, and in many areas dead, sun-bleached stumps are a common feature of the landscape. Only in extremely rocky areas, especially on mountain tops, are living cedars found. In the deeper kloofs rare remnants of high forest remain in the few specimens of rooiels and witels, yellowwood and hard pear, while drier kloof forest, represented by wild olive, silky bark and spoonwood, is more common in the valleys.

Spring, when wild flowers thickly carpet the ground, is the best time to see the Cedarberg. The days are warm, the nights cold but not freezing, and the winter rains have left the streams brimming. The Olifants River valley at this time is flushed with bright yellow and orange daisies, colourful heaths and many unusual endemics. In spring of 1983 Clive and I visited the Cedarberg when the flowers were the finest anyone could remember.

Apart from the cedars, the floral splendour, the ruggedness and remoteness of the mountains, it is the fantastic weathering of the sandstone that makes the Cedarberg unique. Weird and grotesque shapes rise from every rocky outcrop and are silhouetted like chess pieces, spears, armies, fluttering birds and winged gods against the sky. The Valley of the Red Gods below the Wolfberg Cracks,

134. *Greg Lacey leads 'Magnetic Wall' (G1) on Table Mountain. There are a number of ways of getting up the mountain; cable car riders may think the rock climbers are crazy, but they in turn shun the technology that robs people of a mountaineering experience.* **135.** *Aleck McKirdy on 'De Bruin Damage' (G1) demonstrates friction climbing on the granite outcrops of Lion's Head. This is a new technique for local climbers who are more used to the horizontal shelves of sandstone.* **136.** *Selecting and placing protections are vital to a leader's success. Chris Lomax checks his gear, searching for the one 'nut' or 'friend' that will best fit the crack for his running belay on 'Touch and Go' (G1), Table Mountain.* **137.** *The cliffs under the cable car station are known to climbers as Cableway Crag. Beyond them stand the first of the Twelve Apostles.*

139

140

138. Formidable Milner Peak, in the Hex River mountains, for many years defied the challenges of man. **139, 140.** Kloofing has long been an alternative pursuit for mountaineers and hikers. The complex kloofs of the western Cape ranges were first negotiated en route to the climbing areas, but the new sport soon caught on as another adventurous way to explore the mountains.

the Wolfberg Arch and Maltese Cross below Sneeuberg are some of the grander weathered formations that have become familiar to hikers over the decades.

Paths with easy gradients link many scenic spots and, with the exception of the wild Krakadouw region, the whole range can be hiked in a week from north to south without strenuous routes having to be negotiated. The main reason for this is the shale band known as Die Trap, 'The Step', which runs like a shelf at about the 1 400-metre contour beneath the higher peaks of Sneeukop (1 932 metres), Tafelberg (1 971 metres), Groot Krakadouw (1 737 metres) and the highest of all – Sneeuberg (2 028 metres).

The Cedarberg is a climber's paradise, combining great beauty, easy access, good rock and largely unexplored crags. The first climbers to visit Cedarberg were Gother Mann, George Amphlett and Dr Arthur Stark in 1896. They travelled by wagon over the Koue Bokkeveld by way of the Gydo Pass, built 40 years previously by Andrew Geddes Bain.

Tafelberg was their first goal as it had never been climbed before. It was a cold, wet day when Mann and a local coloured man named Viljoen discovered why. They followed a steep route up the gulley behind the Spout column and into a water-worn, cavern-like chimney with polished walls about eight metres high. Ignoring the freezing weather, Mann discarded his boots and stockings and eased his way up the chimney. Viljoen could not follow but a rope, hastily made from a belt, handkerchief and shoelaces, brought him barefoot, wet and very cold to the top of that enchanting mountain. His excitement was so great that he did not notice the cold until he discovered his right hand was completely numb – the thrill of conquering a peak can be overwhelmingly intense.

The next day the party climbed Sneeukop and found a beacon on the summit. In 1843 Astronomer Royal Thomas Maclear (who erected the Maclear beacon on Table Mountain), Thomas Mann (Gother's father) and Wallich had set up a number of stations on Cape peaks to measure the arcs of the meridian, including the one on Sneeukop. From Mann's diary it appears that nothing prevented these astronomers from establishing a station once a peak had been chosen and they suffered great discomfort and hardship in this pursuit. They covered most of the Cedarberg range, often in icy winds and temperatures well below freezing, and proceeded over the eastern limit of the range to Wuppertal where they gratefully exchanged their worn-out shoes for the mission station's famous *velskoene*.

Two years later Thomas Mann was in the mountains near Riviersonderend when he wrote: 'Gales of wind from north with sleet. . . tent ropes thickly encrusted in ice four to five inches thick; wind furious, impossible to face.' Unlike his son, the astronomer did not go to the mountains for recreation, yet was a fine mountaineer who loved the places he visited.

The ascent of the smooth Tafelberg Spout by Frank Berrisford in 1924 opened the way for serious rock climbing in the Cedarberg. Twelve years later J.M. Marcus led a frontal route up Groot Krakadouw, and in 1963 Jan Goedknegt and Binkie Kohler managed a strenuous G route up the Spout while the heavens poured down lightning, thunder and rain upon them. They called it 'Thunder Crag'. The crux pitch had turned back a previous party and now Goedknegt was in trouble. 'My reserves of strength were rapidly diminishing. I had to do something, but quickly. I tried to climb down. I could not. Traversing was out of the question. All that was left was to go up. I tried, but soon retreated to my former uncomfortable situation.

141. *Last light casts a glow on the Wolfberg crags, a fine hiking and rock climbing area in the Cedarberg.* **142.** *Inside one of the Wolfberg Cracks. From the top of the Cracks a rock platform known as Die Trap forms a plateau that runs beneath most of the Cedarberg's peaks and connects many of the most interesting areas in the range.*

'I grew very scared. "I'm stuck," I gulped. "Well knock in a peg," came back up the cliff. I looked around, but there wasn't a suitable crack for the type of pitons I had. I moved up onto the difficult rock again and immediately began to shake so much it felt as if I had contracted Parkinson's disease.

'Never before had I been so scared. I moved up, battling to find the next grip. Ah! there was one, but it sloped the wrong way. I groped around for some more but in vain. I was shaking more than ever as I desperately tried to reach the next grip. I called on every muscle, even the strength of my involuntary organs, to send what power they had to my fingers which wanted to open, but which I kept closed by an effort of will. A few more movements and I would be up.

'I lay exhausted; my fingers closed uncontrollably, my body stopped shaking. After a few moments of bliss I realized how good it was to be alive. I wanted to tell the whole world what a wonderful day it was so I shouted down to Binkie, who replied: "What's so wonderful about it. I'm frozen stiff. Can I climb?"'

Another route was put up the Spout in 1962 by Mamacos and Baillie, again where others had failed. The climb was even harder than expected. 'After long and careful scrutiny the apparently smooth, marble-like rock began to assume suggestive shadows here and there, which turned into faint hollows and finally indicated a possible presence of holes.' Such is route finding.

Climbers seem to thrive on adversity, and a year later Mike Scott was on the Spout in midwinter. 'The snow, which had begun to fall and clog our upward-peering eyes, increased in intensity as we climbed a large block to the right of 'Thunder Crag'. The squelching sound of our socks inside our boots quite unnecessarily told us we had cold feet! When we looked up at the next section they got colder. A few pull-ups landed us in a two-man niche next to a block.

'After a lengthy period had passed in erecting a belay to end all belays, the leader blew on his cold fingers, gripped his kneecaps to stop them wobbling, and climbed a sort of crack by pulling up hand-over-hand on small, incut grips. Eventually a bulging face prevented upward progress.

'The perplexed leader, thinking that the only descent he could possibly attempt under the circumstances would probably give the second man quite a jolt, essayed a remarkably delicate traverse to the right around the bulge. Two thimble grips, into which only one fingertip at a time could be inserted, were sufficient to enable him to retain purchase on the rock. After moving up, a mantleshelf onto a precarious ledge under a large overhang was accomplished by using an underhand grip in order to stand up. Two or three pitons were inserted at this juncture, for the next pitch required strenuous and alarming stretching.'

For the past two decades the Cedarberg has been a great pioneering ground for rock climbers and an increasingly popular wilderness for hikers. A number of climbing routes in the Cedarberg, namely at Krakadouw and Tafelberg, feature in Dave Cheesmond's booklet on 50 great climbs of the Cape.

From 1977 the Wolfberg Cracks suddenly attracted the rock 'jackals' from Cape Town who opened 'Red Fox', 'Lone Wolf' and 'Sheep's Clothing' before the cry went out. André Schoon began the attack with the 11-pitch traverse 'Wolf in Sheep's Clothing'. In 1978, within just six weeks, a dozen new routes from F1 to G3 were pioneered, including two G3 aided routes, 'Celestial Journey' and 'Alone in Space', which were opened by David Davies and Robin Barley and later free climbed.

At present there are 22 bodies in the country to control nature conservation. The Cedarberg is currently under the careful protection of the Department of Forestry, as a rain catchment area. Now there is talk of turning it into a national park under the centralized control of the National Parks Board. While the rationalization of conservation is a sensible idea, mountaineers and conservationists alike fear that any change in the control of the Cedarberg could turn this mountain wilderness into a commercialized tourist spot. The thoughts on conservation expressed by D.P. Ackerman, a former Secretary for Forestry, should be heeded: 'We need wilderness areas for their scientific interest as natural ecosystems, we need them for their aesthetic value as unscarred landscapes, we need them for the opportunities they offer in spiritual and physical recreation, we need them just for the sake of knowing that they are there – untouched by development of any kind.'

143

143. A dead cedar tree is reminiscent of the forests that grew in the Cedarberg before generations of woodcutters exploited the excellent wood for the fast-growing Cape Colony.   144. Shallow pools lie in the water-eroded depressions on Tafelberg. On the horizon Sneeuberg, at 2 028 metres, is the highest peak in the Cedarberg.   145. The route up Sneeuberg from Dwarsrivier passes the monolithic Maltese Cross on a stony plateau.

146. Seen from Welbedacht Cave in early spring, Tafelberg and the Spout catch the last gloomy rays of the sun.   147. (Overleaf) Scattered cedar trees, dead and living, stand among the clinkered fragments of rock in the Cedarberg. Wind and water have created fantastic rock formations which make these mountains one of the most impressive wilderness areas in southern Africa.

146

NAMIBIA
AFRICA'S FORGOTTEN PLAYGROUND

As the wasteland is tempered by
  the sun's furnace fire
Afternoon clouds grow lazily,
  higher;
Rhythmic, brooding, swelling to burst
Upon the country's unquenchable thirst.
A tale of genesis unfolds 'cross the land,
A Garden of Eden, still smothered
  by sand.

Ovambo Interlude

To the eye accustomed to wooded peaks, fresh mountain streams and the varied landscapes of other highland areas in southern Africa, the austerity of line and colour in the mountains of Namibia can be forbidding. From these pristine ranges unfolds an intensity of experience beside which the abundance of the Drakensberg or Cape Fold mountains seems overstated. For Namibia is a harsh paradise, its central plateau squeezed between two deserts. The land is wide and arid, and when the infrequent rains do come, they are usually in sudden, violent storms.

Across the dry plain you may hear the moan of the wind, its easterly gales flinging sheets of hot, stinging sand towards the sea – or nothing at all. A great stillness pervades this desert where little moves, little grows. And it is on the fringes of the Namib that the more interesting mountain ranges lie – the Pontoks and Spitzkoppe, the Erongos and Brandberg. Concentrated mainly between Windhoek and the Skeleton Coast, they lie between unforgiving desert and dry thorn savannah, rising sharply out of the flat surroundings. The Pontoks and Spitzkoppe are granite cones that look like giant eggs pointing upwards, their bases buried in the sand. The Erongos, lying between the Brandberg massif and Pontok range, form a wavy line of granite

148. *Eckhard Haber, Errol Nienaber and Clive Ward made the first ascent of the north face of Spitzkoppe (G1, M1) in 1981.* **149.** *Wind-sculpted granite outcrops and boulders near Spitzkoppe, looking across to the Sugarloaf.*

hills with bizarre rock formations sculptured by the wind.

The Namibian mountains are a jumble of rock, resulting from nearly all the geological processes we have already witnessed at work in South Africa. Protruding from the desert sands are outcrops of primitive granite, limestone, marble, banded ironstones and other ancient deposits of Precambrian times. Vast Karoo sediments, mainly to the north in the Kaokoveld, still lie beneath the

lavas that poured across southern Africa in rivers and lakes of fire.

This scarred landscape is, perhaps, a fitting monument to its past, for it has been rent by more volcanic explosions and intrusions than any other area in southern Africa. Mount Brukkaros, rising 550 metres above the empty plain north of Keetmanshoop, is the result of one such explosion. This now-extinct volcano formed when a kimberlite pipe could not reach the surface and, literally, 'blew its top'. As the molten magma collided against the sediments of the earth's crust, the explosion must have been spectacular for it left a crater 2,4 kilometres in diameter.

After the period of fire came the great ice sheets that froze the land for millions of years, just when higher life forms were struggling to establish themselves. As the sedimentary deposits left by the melting ice were worn away the intrusive volcanic rocks were slowly exposed, and convex granite hills rose from the eroded landscape.

Despite their remoteness, the Namibian ranges are gradually becoming familiar to mountaineers, although much is still to be learned of their ecology and natural history. Probably the least visited mountain area is the upper reaches of the Brandberg, yet this massif is the most impressive of all. Its name has many connotations but no-one is certain of the origin: perhaps from an early explorer called Brand, or the burnished cuff of this volcanic fist, or the scorched desert landscape upon which it stands.

It is little wonder that the peaks have hardly been explored for they are among the highest on the subcontinent and, apart from the logistics of getting there, water is scarce on the Brandberg and few people know where to find the occasional

permanent pools, most of which lie hidden in crevices. Furthermore, the heat is relentless, and any climbers on the main peaks would have to carry about five litres of liquid a day and be sure to find a sheltering overhang under which to escape the midday sun.

A German officer, H. Jochmann, was the first European known to have explored the Brandberg. He visited the Tsisab valley in 1909 and discovered the first of many rock paintings there. It is not known how long ago the mountains were inhabited, nor for how long, but the paintings suggest that the land was once kinder and big game more abundant. After the Drakensberg, the Brandberg has the most impressive collection of rock paintings in the world; nearly every visit by keen observers extends our knowledge of this primitive art gallery.

The first known attempt to climb the Brandberg was by two surveyors, Burfeindt and Carstensen, in 1914. Extracts from one of their diaries reveal both the severity of their task and their determination:

'Accompanied by one European and four natives, six saddle horses and four mules, we rode all along the steep slopes of the Brandberg until we reached the dry Ugab river.

'The discovery of water was accidental, but relieved us of our most pressing worries. The accompanying natives were christened in the Numas gorge from which a successful attempt was also made on the Königstein, the highest peak. Hunger, thirst and despair had to be endured. Members of the expedition lived on mice and the sinewy meat of leopards. Sudden weaknesses and sickening of men and beast occurred.

'On receiving the tragic news of the outbreak of the First World War, the expensive surveying equipment was carefully stowed behind a rock face and the exhausting homeward trek attacked.'

Some records suggest that Burfeindt and Carstensen really climbed the fourth highest peak, the Horn. The Königstein – the highest peak in Namibia, and the highest in southern Africa outside the Drakensberg massif – was probably climbed only in January 1918 by another team of German surveyors. One of this later party, Dr Reinhard Maack, was a meticulous naturalist and in 1923 published a report on the geography and geology of the area, including some fascinating notes on the vegetation. He was the first to describe the graceful Brandberg acacia and the Brandberg euphorbia, two of the most striking plants of the range.

The most startling discovery of the expedition, however, was of a different kind. Having measured the peak at between 2 606 and 2 614 metres (its height has since been fixed at 2 697 metres), the party stayed near the top of Königstein until food and water ran out. The descent was difficult and the heat overpowering. At one stage Maack sought refuge from the sun under a rock

shelter and fell asleep. When he awoke his eyes focussed on a fantastic rock painting of a group of people gathered around a white figure. He made a rough sketch of his find, identifying the painting as a fresco of Egyptian style. It is uncertain who painted it – Bushmen, the negroid Bergnama or some earlier, possibly more sophisticated people. The famous archaeologist Abbé Breuil visited the Tsisab ravine in 1947 and his publications on the Brandberg paintings attracted worldwide attention. Despite current belief that the figure portrayed is that of a young man, the painting is still known by the name accorded it by Breuil, the 'White Lady of the Brandberg'.

The second ascent of the Königstein was made only in 1943, by Cape mountaineer Denis Woods. He found more rock paintings, grouped around a spring where long grass, reeds and wild fig trees break the harshness of the burning mountains; the rocky waterfall must have been an important water

supply for the people who lived or sheltered there.

The Brandberg's great complex of ridges, ravines and terraces conceal a number of rock art treasures that only the most determined climbers ever see. On the upper Numas plateau the so-called 'Okapi' painting depicts a pregnant hartebeest at the centre of a hunt. Another difficult climb over terrace after terrace ends on a wide sandy platform encircled by large boulders – the Felsencircus. Here on an overhang is a mural of monochrome and polychrome rock paintings of gracefully leaping springbok, crane-like birds, and hunters.

Scientists delight in the Brandberg's slowly emerging mysteries. The archaeological interest in the area attracted a large expedition in 1955 and during its three-week stay the party was resupplied by air drops. A member of the expedition, H.J. Wiss, recorded a number of new plants, among them *Mentha wissii* and *Plumbago wissii*. Botanist Bertil

150. *The domes of the Pontok mountains, with the Pontok Spitz in the centre and Pontoks 1 to 4 to the left, offer a number of high-quality rock climbs.* **151.** *(Overleaf) Spitzkoppe from the north-east, as sunset casts long shadows across the desert.*

Nordenstam has made numerous journeys there over the years. From his diary we can retrace his progress: 'First visit – 6 May. Climbing upwards, a better waterplace with a large sycamore tree was found. A faded giraffe painting was noted on a cliff wall. The valley narrowed to a steep and difficult gorge and the ascent was continued on the western slope.

'The flora was rich, with trees of *Cyphostemma currori* [cobas trees] now starting to occur and a single kokerboom was observed on the ridge. At the junction of a side valley with the Tsisab, a wonderful system of waterholes [the '1 200-Meter Wasser'] was found. One of the rock pools, three metres across and as deep as a man, was especially suited for

swimming and we named it Flensburg's Bad. Returned to last camp.'

Archaeology here is hampered by the same extreme conditions that hamper mountaineering, and most of what is known about the range has been deduced from material collected around a few sites at its base where, after periodic rains, water gushing down steep ravines feeds a few waterholes. At the entrance to the Tsisab ravine crude quartz tools from the Late Stone Age lie among the masses of collapsed basalt that look like heaps of crumbled rock, the remains of a long-forgotten stone city.

The curator of archaeology at the State Museum in Windhoek, J. Kinahan, notes that in the imaginations of its European conquerers the Brandberg became 'a world apart'. Their fantasy of early Mediterranean traders in the Brandberg caused a distorted view of the mountain's past. Despite the 'White Lady's' Cleopatra-style hair, contact between the Brandberg and Mediterranean peoples has now been completely discredited and the African origin of the painting is no longer in doubt. The prehistory of this area is similar to that of the rest of the subcontinent – it never was a world apart.

Recent surveys of more than 80 archaeological sites in the Hungarob valley, on the south-western side of the mountain, are expected to lead to a more complete understanding of the massif's early inhabitants. At least 8 000 years ago the Brandberg was frequented by Late Stone Age hunter-gatherers who fashioned implements from the quartz and basalt in the area and used these tools to make their simple equipment. They tanned skins for clothing and bags, made wooden bowls and utensils, spear shafts and bows, tipped their arrows with stone barbs and used bone link-shafts to attach

se points to the arrows. Caves and
ne shelters near reliable waterholes
came their dwellings, but their
idence at any particular site was
iited: nomadic bands probably visited
 same sites over and over as they
lowed the seasonal flux.
The higher waterholes supported small
nds for longer than those lower down,
d this often resulted in a dilemma: to
ove out and hunt the game on the desert
ains after the rains had ceased, or to
 age for as long as possible at the
terholes. Both options entailed risks,
t mobility and timing were key factors.

153

152. *At the head of the dry Tsisab valley in the Brandberg, Königstein (2 697 metres) is the highest point in Namibia. The 'White Lady' frieze is painted on an overhang in this valley.*
153. *The 'White Lady' of the Brandberg was first thought to be of Mediterranean origin but is now believed to be an unusual variation of indigenous rock art, portraying a negroid figure with white body-paint.*

About 2 000 years ago the hunter-gatherers began to keep small stock, probably sheep and goats, and although their essential requirements did not change, the herding of stock almost certainly led to a slow transformation of the social order. Accumulated wealth represented by domestic animals undermined the simple life style of the nomadic bands, although evidence of very few changes during the initial phase of stock herding suggests that the ecology was flexible enough to absorb this means of subsistence as well as hunting. However, the Brandberg and surrounding areas became increasingly dry and, as the nomads and their herds were less mobile and more dependent on water, their decisions on where and when to move would have been drastically influenced.

The mountain probably became one of a number of options in the seasonal round of pastures and waterholes. When the first Europeans visited the Brandberg towards the end of the nineteenth century they met large groups of pastoral nomads who were camped around the base of the mountain. Until the 1940s these nomads moved freely across the arid Damaraland plains. Then, for a short time, they were anchored to government-subsidized boreholes around the mountain, where water was more easily available. However, the break in the age-old migration patterns upset the delicate balance between naturally available water

and grazing, resulting in over-use of the land. Today the Brandberg lies deserted while the Damara herdsmen's flocks feed in the thornveld along the Omaruru River to the east.

Finding permanent water in the Brandberg may be a problem for transient humans, but the Rüppell's parrots which chatter at the top of acacias, the few zebras which still roam its lower terraces and the leopard that coughs and roars after a night of successful hunting know where it is.

From the watered basins between the Königstein and Numasfelsen, the highest peaks of the Brandberg, plunge the steep, boulder-strewn gulleys of the Messum valley. Opening out at the base of these 'burned mountains', the valley drifts out into the desert with only water-rounded pebbles and large anaboom trees to show where the river should flow.

Looking down the Messum valley from the Königstein, the Namib Desert blends with the sky in a fiery haze on the horizon. With the Orange River as its southern border, it runs the entire length of the country, through the northern reaches where it is known as the Kaokoveld, and into southern Angola where it is the desert of Mossamedes. It is like the sandpit of some giant-child, where empty vistas distort all proportion: the huge dunes are like sand ripples, the massive granite outcrops and boulders that litter the endless gravel like the giant's long-forgotten marbles.

154

155

154. *Clive Ward climbs up to the bivouac on the first ascent of the 'South-west Wall' on Spitzkoppe. Across the plain is Klein Spitzkoppe.* **155.** *A party abseils off Spitzkoppe after climbing the 1947 route, now known as 'Standard Route'.* **156.** *The west face of Spitzkoppe is the most daunting rock slab in Namibia. The route 'South-west Wall' follows the diagonal crack up the right-hand side of the wall.*

156

During the day the heat is vicious, while night-time temperatures may plummet below zero. Perhaps the hand of nature forgot to clothe this remote tract of Africa, or in a fit of anger stripped it bare to suffer intense heat, cold and the constant lashing of wind-whipped sands. But a closer look reveals a fascinating world between the grains of sand, where every flower and insect is an evolutionary masterpiece. All is patient and finely attuned to the land's slow, irregular cycle; it seems that here eternity can indeed be held in the palm of one's hand.

Evolution has equipped many creatures and plants to live here. When moist air

sand. On its long hind legs it elevates its body so that the carapace becomes a shield on which the moisture condenses, and is then channelled down into the beetle's mouth. Some *Onymacris* species have white backs that reflect the sun's radiant heat, thus allowing the beetles to spend more time foraging for food above ground than do their black-backed counterparts. According to popular belief, one of the white-backed beetles climbs onto another's back to provide shade while its partner forages. It has been established, however, that its motives are not so altruistic, for the beetles are in fact copulating.

themselves in a side-winding wave motion that gives them maximum traction with minimum surface contact.

A bright moon reveals nocturnal creatures leaving their daytime shelters to search for their next meal. Termites, crickets and many strangely shaped beetles emerge and are attracted by the high-pitched, barking call of the translucent palmate gecko, which then gobbles them up. The small, web-footed gecko makes a meal for the white dancing spider which sleeps during the day in a web-lined tube. The size of a large human hand, the spider hops around aggressively when hunting and, if it needs to make a

157

from over the cold Benguela Current makes contact with the hot land air, thick blankets of mist roll over the desert and condense on the ground to sustain nearly all the life of the Namib. It has recently been discovered that the Namib is one of the world's great lichen belts, and that the wind-scattered lichens provide food for most of the desert creatures.

When the mists blow over the dunes, the beetle *Onymacris unguicularis* emerges from its granular cavern in the

Other creatures of the Namib, such as the sand runner beetle, have unusually long legs to keep their bodies well above the baking ground. To help regulate its body temperature, the lizard *Aporosaura* does a quickstep shuffle to keep as few feet as possible on the ground at once. Desert snakes around the world have adapted in similar ways to cope with the hot, loose sand. In the Namib, Peringuey adders lie under the sand in wait for prey and when they decide to move, propel

sudden departure, curls up its legs and rolls down the dune. Other spiders make sails out their bodies and allow the wind to carry them away. Shovel-nosed lizards eat the spiders and are in turn preyed upon by adders.

Vegetation, too, has adapted to cope with the adverse conditions. In the broad, flat channels of the Pro-Namib, a transition zone between the desert and the mountains where little else can survive, *Welwitschia mirabilis* grow. They are

most common around the Brandberg but also occur scattered throughout the northern Namib. The plant's two leathery leaves persist for its entire life, and would reach a considerable length were they not continually shredded by the elements. Its fibrous trunk looks like a chunk of driftwood that is slowly succumbing to the shifting sand, and is surrounded by long streamers of leaf. The tap root penetrates deep into the dry earth, but a meagre water supply is also provided by droplets of moisture that condense on the leaves and drip to the ground, to be greedily absorbed by a network of fine surface roots. When in flower, the plant is covered with dark red spikes like red-gloved fingers groping at the dry air. Although welwitschias are believed to live for about two thousand years, most of those seen are estimated to be a youthful few hundred years old.

A journey through this country is like travelling through a dream landscape in which the vegetation no longer seems to be as vegetation should, for the extreme arid conditions of the Namib have produced a number of other botanical oddities. Two examples are the knobbly succulent *Hoodia gordonii*, known as queen of the Namib, and *Jensenobotrya lossowiana*, that looks like an enormous bunch of grapes left behind by a hurried traveller. A large-leaved species of *Nicotiana*, similar to the tobacco plant, has recently been recorded and is very rare. Succulents with thick, water-storing stems, such as the cobas tree, grow among rocky outcrops around the Erongo and Pontok mountains. The tall *Moringa ovalifolia*, commonly known as the upside-down or phantom tree, is believed by the Bushmen to have fallen head-first from heaven, as its bulbous trunk narrows considerably to the crown and the

157. *Growing among the boulders of Spitzkoppe, this papery-barked cobas tree is one of several plants that have adapted to the particularly harsh environment of the Pro-Namib.* **158.** *Like a bonsai in a rock garden, this tree sucks water from deep in the cracks to survive on the fringes of the Namib Desert.* **159.** *Welwitschia bugs (Odontopus sexpunctatus) forage on the cones of a Welwitschia mirabilis, the strangest plant of the Namib. The red bug is the larval form of the yellow Odontopus.* **160.** *A gecko blends so well into the granite of the Pontoks it seems almost to emerge from the rock, like one of the organic reptilian embryos that were once believed to be the origin of fossils.* **161.** *Cool springs on top of the Brandberg, between the peaks of the Königstein and Numas Felsen, create marshy areas which nurture water-loving plants such as Androcymbium melanthioides.*

branches that sprout at the top of the tree look like roots.

The Namibian escarpment is not always as well defined as that of the Transvaal and Natal, but its broken step can be traced from the Richtersveld south of the Orange River right up to Angola, and is at its most prominent between Windhoek and Swakopmund where the mountains are called the Khomas-Hochland. To the south-west, in the centre of the Namib Desert, lies the Naukluft range, its orange hills broken by a series of black dykes flowing like waves along its entire length. East of the escarpment, grassy savannah muscles between the Namib and Kalahari, while from the interior plateau dry riverbeds follow winding gorges for about 50 kilometres through the mountain barrier, connecting the two worlds.

The Germans who first colonized South West Africa were interested in neither the land's essential beauty nor its potential wealth, for its mineral treasures had not yet been discovered. This barren chunk of Africa was simply the last piece of the continent not snatched by other European empire-builders, its only visible advantage being that it offered ships a well-protected harbour at Lüderitz Bay. With Teutonic thoroughness they measured the country, from the mountain peaks to the shifting dunes, with cumbersome chains. If the English have been likened to a nation of shopkeepers, then surely the Germans must be a nation of surveyors, for there was little of their new colony they did not measure.

The German surveyors climbed the Brandberg as faithful servants of their country, but it was South African Denis Woods who was first drawn by the mountains' strange beauty. He explored many remote Namibian mountains, opening the way for later climbers. In 1940 he made two attempts at scaling Spitzkoppe (1 759 metres) via the southern ridge but was finally defeated by the 'Gendarme' rock, a mere hundred metres from the summit. To his companion I.C. Smith, Spitzkoppe looked like the tip of a giant elephant's tusk. 'Some mountains are friendly, this one was not,' he wrote. 'For me it had an air of hostility. The scanty vegetation of its slopes was either prickly or poisonous.' That is so, but today the mountain presents climbers with more of an adventure than a problem, and wandering around its perimeter is like taking a walk through a geological amusement park.

In 1945 a party of German climbers from Windhoek was turned back by the same rock which had thwarted Woods and Smith. 'Shippy' Shipley and his party tried the southern ridge route in 1946, were defeated and pioneered instead the north-east side of Spitzkoppe. Intuitive route finding led them through granite labyrinths, up a number of long chimneys, through dark and winding 'worm-holes'. About two-thirds of the way up the mountain they had to abseil over a ledge – only to be beaten by a smooth wall of rock five pitches from the top.

Four days later they returned and defied all climbing ethics by chiselling steps into the blank section. For seven days they battled on the face, but in the end were driven back by a gale. Although they found a way up to the summit, they did not make the first complete ascent. Even the highly respected team of Jannie Graaff and Hans and Else Wong were forced to retreat several times before they finally reached the summit in 1947.

Wind erosion is a strong factor in shaping the Namibian landscape. Giving the final polish to the mountains, gales sandblast the rock and carry away most of the loose material to leave smooth cones,

balls of rock and top-heavy pedestals. The Spitzkoppe is one such towering cone of blank granite. Its 550 metre-high south-west face is the largest and smoothest rock face in the country, and was first climbed by Clive Ward and Eckhard Haber in 1982, after numerous parties had been 'psyched out' by the exposure experienced while scaling its longest crack. Clive and Eckhard spent three days on this formidable wall, the rough granite crystals chafing their limbs like an unrelenting cheese grater. Some sections were so devoid of holds or protections that they had to drill small holes in the rock and climb with the aid of bathooks. Their route was followed in 1983 by two parties from the University of the Witwatersrand.

The neighbouring Pontoks are more easily ascended, but some difficult rock climbing routes have been established on them. Their name is a corruption of 'pondok', the Afrikaans term for the ramshackle hut that each of the four main domes resembles. From a distance they look like the soft rumples of an unmade

bed, but closer inspection reveals the coarse decomposing granite that will grate the skin of anyone unlucky enough to lose his grip on the rock.

Climber Walter Friedrich writes of his trip there: 'Beyond Usakos we crossed the arid plains towards the towering silhouettes of the famous Great Spitzkop and the Pontok mountains, dome-shaped peaks which stood like guiding landmarks on our journey. . . Behind us in the darkness the beautiful monument of earth's evolution rose to partly obscure the clear and starry sky, until the rising sun capped the distant Erongo range and made Spitzkop glow like a Chinese lantern casting red and gold reflections across the plains.'

While climbing the Klein Spitzkop Friedrich and his companions saw the mauve flowers of the Bushman's candle for the first time and 'against all common belief' a welwitschia bug on a resurrection plant. As its name suggests, this insect's life cycle is inextricably linked with the welwitschia, a plant that does not occur

on the Klein Spitzkop. The find of the day, however, was a fair-sized silver topaz among the many quartz crystals scattered around. Indeed, the ground around the Spitzkoppe has long been known to contain a rich supply of semi-precious stones, including topaz, aquamarine and tourmaline. These mountains, too, boast collections of Bushman paintings on overhangs and in hidden caverns, and even today a keen observer may discover Stone Age artefacts.

A short distance across the desert, to the north-east of the Spitzkoppe, lie the Erongo mountains. This ring of granite and volcanic rock was visited by Denis Woods during his early explorations of Namibia as well as by later travellers who delighted in its Bushman paintings, of which the White Elephant is best known.

Helmut Brass, a 'South-Wester' who visited the Erongos frequently during the 1960s and 1970s, described spectacular giant boulders, some standing on delicate-looking pedestals, and said of the area: 'Farm tracks are very few, and due to the inaccessibility of some parts a large area of the mountain is still unexplored and unspoiled, and the lush vegetation has not yet given way to haak-en-steek, steekgras and duiweltjies.'

The early 1980s were some of the driest years recorded in Damaraland and contrast with Brass's experiences there: 'The small streams combine on a high plateau to become rivers rushing down over enormous granite banks into narrow rocky channels, diverted into many channels by rocks and boulders buried among the rich vegetation, and plunging again over bare rocks to reach a lovely, lonely valley.' Such were the climatic variations that dictated the movements and lives of the hunters and herders who made this harsh land their home so many years ago, and who left only their art to tell us who they were.

For a million years and more the mountains have been our hearth and if in a million years there are still humans on this planet, then it is the mountains that will shelter their tired spirits. In the ancient hills of the Transvaal we have discovered our beginnings: here among the Omukuruwaro mountains, the 'burned mountains' of Namibia, we see an end – or perhaps just a new beginning.

162. *From Spitzkoppe dry plains reach to the basalt-crested Erongo mountains, the remnants of an ancient volcano.*
163. *Climbers pick their way down the shattered scree of the east ridge after climbing a route on Ababis mountain, a massive quartz outcrop to the south of Spitzkoppe.*
164. (Overleaf) *Hottentots Cathedral was named after a group of bandits who descended from their mountain hideout to raid wagon trains as they stopped at the watering hole at Wilsonsfontein on their long journey between coast and hinterland.*

REFERENCES

Journals of the Mountain Club of South Africa, Cape Town Section (annual).

Journals of the Witwatersrand University Mountain Club, University of the Witwatersrand, Johannesburg (annual).

Veld and Flora: Journal of the Botanical Society of South Africa. Volume 70, number 2, 1984.

Acocks, J.P. 'Veld Types of South Africa'. *Memoirs of the Botanical Survey of South Africa*, number 40, Department of Agriculture, Pretoria. 1975.

Becker, P. *Path of Blood: the rise and conquests of Mzilikazi, founder of the Matabele nation*. Penguin, London. 1962.

Brandt, H. *Backpackers' Africa*, Brandt Enterprises, Cambridge. 1983.

Burman, J. *A Peak to Climb*. C. Struik, Cape Town. 1966.

Chouinard, Y. *Climbing Ice*. Hodder and Stoughton/Sierra Club Books, San Francisco. 1978.

Clarke, J. and Coulson, D. *Mountain Odyssey in Southern Africa*. MacMillan, Johannesburg. 1983.

Cluver, M.A. *Fossil Reptiles of the South African Karoo*. South African Museum, Cape Town. 1978.

Day, J., Siegfried, W.R., Louw, G.N. and Jarman, M.L., eds. *Fynbos Ecology: A Preliminary Synthesis*. SA National Scientific Programmes Report number 40, CSIR. 1979.

Dickson, J.R. and Wannenburgh, A. *The Natural Wonder of Southern Africa*. C. Struik, Cape Town. 1984.

Frandsen, J. *Birds of the South Western Cape*. Sable Publishers, Johannesburg. 1982.

Grobler, H. and Hall-Martin, A. *A Field Guide to the Mountain Zebra National Park*. National Parks Board, Pretoria. 1982.

Hall, A.V., De Winter, M., De Winter, B. and Van Oosterhout, A.V. *Threatened Plants of Southern Africa*. SA National Scientific Programmes Report number 45, CSIR. 1980.

Hamilton, G. and Cooke, H. *Geology for South African Students*. Central News Agency, Johannesburg. 1954.

Irwin, P., Ackhurst, J. and Irwin, D. *A Field Guide to the Natal Drakensberg*. Wildlife Society of Southern Africa. 1980.

Kinahan, J. 'Walking to the Rain: Prehistoric Nomads of the Brandberg'. *SWA 1983*. SWA Publications, Windhoek. 1983.

Kitching, J.W. 'The Distribution of the Karoo Vertebrate Fauna'. *Memoir of the Bernard Price Institute for Palaeontological Research*. 1977.

Levy, J. *Everyone's Guide to Trailing and Mountaineering in Southern Africa*. C. Struik, Cape Town. 1982.

Luckhoff, H.A. *The Sederberg Wilderness Area*. Department of Forestry and Environmental Conservation, Pretoria. 1980.

Marais, E.N. *My Friends the Baboons*. Human and Rousseau, Cape Town. 1971.

Marais, E.N. *The Road to Waterberg and other essays*. Human and Rousseau, Cape Town. 1972.

McLachlan, G. and Liversidge, R. *Roberts Birds of South Africa* (revised edition). John Voelcker Bird Book Fund, Cape Town. 1982.

Messner, R. *Solo Nanga Parbat*. Oxford University Press, New York. 1980.

Miller, P. *Myths and Legends of Southern Africa*. T.V. Bulpin Publications, Cape Town. 1979.

Mountain, E.H. *Geology of Southern Africa*. Books of Africa, Cape Town. 1969.

Murray, J., ed. *Cultural Atlas of Africa*. Phaidon Press, Oxford. 1981.

Pearse, M.L. *A Camera in Quathlamba*. Howard Timmins, Cape Town. 1980.

Pearse, R.O. *Barrier of Spears*. Howard Timmins, Cape Town. 1982.

Rousseau, L. *The Dark Stream: the story of Eugène N. Marais*. Jonathan Ball, Johannesburg. 1982.

Skaife, S.H. *African Insect Life* (revised edition). C. Struik, Cape Town. 1979.

Steyn, P. *Birds of Prey of Southern Africa*. David Philip, Cape Town. 1982.

Trauseld, W.R. *Wild Flowers of the Natal Drakensberg*. Purnell, Cape Town. 1969.

Truswell, J.F. *The Geological Evolution of South Africa*. Purnell, Cape Town. 1977.

Van der Post, L. *The Heart of the Hunter*. Penguin, London. 1965.

White, F. *The Vegetation of Africa*. UNESCO. 1983.

Willcox, A.R. *The Rock Art of South Africa*. Nelson, London. 1963.

Wilson, K., ed. *The Games Climbers Play*. Sierra Club Books, San Francisco. 1978.

INDEX

65. *A geological amusement park lies around the base of Spitzkoppe and the Pontoks.*